Multimodal Biometric Systems

Internet of Everything (IoE): Security and Privacy Paradigm

Series Editors:
Vijender Kumar Solanki, Raghvendra Kumar, and Le Hoang Son

Data Security in Internet of Things Based RFID and
WSN Systems Applications
Edited by Rohit Sharma, Rajendra Prasad Mahapatra, and Korhan Cengiz

Securing IoT and Big Data
Next Generation Intelligence
Edited by Vijayalakshmi Saravanan, Anpalagan Alagan,
T. Poongodi, and Firoz Khan

Distributed Artificial Intelligence
A Modern Approach
Edited by Satya Prakash Yadav, Dharmendra Prasad Mahato,
and Nguyen Thi Dieu Linh

Security and Trust Issues in Internet of Things
Blockchain to the Rescue
Edited by Sudhir Kumar Sharma, Bharat Bhushan, and Bhuvan Unhelkar

Internet of Medical Things
Paradigm of Wearable Devices
Edited by Manuel N. Cardona, Vijender Kumar Solanki, and Cecilia García Cena

Integration of WSNs into Internet of Things
A Security Perspective
Edited by Sudhir Kumar Sharma, Bharat Bhushan, Raghvendra Kumar,
Aditya Khamparia, and Narayan C. Debnath

IoT Applications, Security Threats, and Countermeasures
Edited by Padmalaya Nayak, Niranjan Ray, and P. Ravichandran

Multimodal Biometric Systems
Security and Applications
Edited by Rashmi Gupta and Manju Khari

For more information about this series, please visit: https://www.routledge.
com/Internet-of-Everything-IoE/book-series/CRCIOESPP

Multimodal Biometric Systems
Security and Applications

Edited by
Prof. (Dr.) Rashmi Gupta and
Dr. Manju Khari

CRC Press
Taylor & Francis Group
Boca Raton London New York

CRC Press is an imprint of the
Taylor & Francis Group, an **informa** business

MATLAB® is a trademark of The MathWorks, Inc. and is used with permission. The MathWorks does not warrant the accuracy of the text or exercises in this book. This book's use or discussion of MATLAB® software or related products does not constitute endorsement or sponsorship by The MathWorks of a particular pedagogical approach or particular use of the MATLAB® software.

First edition published 2022
by CRC Press
6000 Broken Sound Parkway NW, Suite 300, Boca Raton, FL 33487-2742

and by CRC Press
2 Park Square, Milton Park, Abingdon, Oxon, OX14 4RN

© 2022 selection and editorial matter, Rashmi Gupta and Manju Khari; individual chapters, the contributors

CRC Press is an imprint of Taylor & Francis Group, LLC

Library of Congress Cataloging-in-Publication Data
Names: Gupta, Rashmi, editor. | Khari, Manju, editor.
Title: Multimodal biometric systems: security and applications / edited by Rashmi Gupta and Manju Khari.
Description: First edition. | Boca Raton, FL: CRC Press, 2022. |
Series: Internet of everything (IoE) | Includes index.
Identifiers: LCCN 2021015970 (print) | LCCN 2021015971 (ebook) |
ISBN 9780367685560 (hardback) | ISBN 9780367685577 (paperback) |
ISBN 9781003138068 (ebook)
Subjects: LCSH: Biometric identification.
Classification: LCC TK7882.B56 .M87 2022 (print) | LCC TK7882.B56 (ebook) |
DDC 006.2/48—dc23
LC record available at https://lccn.loc.gov/2021015970
LC ebook record available at https://lccn.loc.gov/2021015971

ISBN: 978-0-367-68556-0 (hbk)
ISBN: 978-0-367-68557-7 (pbk)
ISBN: 978-1-003-13806-8 (ebk)

DOI: 10.1201/9781003138068

Typeset in Times
by codeMantra

Contents

Preface

This book describes the most recent innovations and technologies that have been introduced during the past two decades for multimodal biometric recognition and its security issues and applications.

The need for biometrics frameworks has expanded massively in commonplace activities such as airport security, medical diagnostics, ATM security, determination of parentage, border security, securing electronic data, E-commerce, online banking transactions, cellular phones, national ID cards, driving licences, the identification of corpses, the investigation of crimes and several other fields.

The novel techniques for biometric frameworks are advancing quickly and boosting the expansion of the technology into new areas. In biometrics the term "multimodal" is used to combine two or more different Biometric sources of a person (for example, facial recognition and fingerprinting) sensed by different sensors. Two different properties (like infrared and reflected light of the same biometric source, 3D shape and reflected light of the same source sensed by the same sensor) of the same biometric can also be combined. In orthogonal multimodal biometrics, different biometrics (for example, facial recognition and fingerprinting) are involved with little or no interaction between the individual biometric. Independent multimodal biometrics process individual biometric independently.

Multimodal Biometric Systems: Security and Applications gives up-and-coming surveys of insights, methods and speculations utilized as a part of biometric innovations for identification and their applications. Specifically, the far-reaching potential of various biometric works and implementations are shown to be a significant array of tools for use by biometric engineers and organizations.

Specialists, researchers, graduate students, designers, experts and engineers who work with biometric research and security related issues will find much of interest in these pages. The material has been organized into independent chapters to provide readers great readability, adaptability and flexibility.

MATLAB® is a registered trademark of The MathWorks, Inc. For product information,

please contact:
The MathWorks, Inc.
3 Apple Hill Drive
Natick, MA 01760-2098, USA
Tel: 508-647-7000
Fax: 508-647-7001
E-mail: info@mathworks.com
Web: www.mathworks.com

Editors' Biographies

Prof. (Dr.) Rashmi Gupta is a Professor in the Electronics and Communication Engineering Department, Netaji Subhas Institute of Technology (Formerly Ambedkar Institute of Advanced Communication Technologies and Research), Govt. of NCT of Delhi, India. She is also holding a position of Founder Director of Incubation Center AIACTR-IRF funded by Delhi Govt. Prior to this she worked in industry for eight years and also has nineteen years of teaching experience. She earned her M.E. and Ph.D. degrees in Electronics and Communication Engineering from Delhi College of Engineering, Delhi University. She has also organized International conference sessions, Faculty development Programmes, workshops and industrial meet in her experience. She delivered expert talks in International Conferences and Faculty Development Programs. Prof. Gupta is associated with many International research organizations as editorial board member and reviewer. She has authored over 85 research papers in various renowned international journal and conferences. She is Fellow member of IETE and senior member of IEEE. Her primary research interests are machine learning, computer vision, signal and image processing.

Dr. Manju Khari is an Associate Professor in Jawaharlal Nehru University, New Delhi, prior to this university she worked with Netaji Subhas University of Technology, East Campus, formerly Ambedkar Institute of Advanced Communication Technology and Research, Under Govt. of NCT Delhi. Her Ph.D. in Computer Science and Engineering from National Institute of Technology, Patna, and she received her master's degree in Information Security. She has 80 published papers in refereed National/International Journals and Conferences (viz. IEEE, ACM, Springer, Inderscience, and Elsevier), 10 book chapters in a Springer, CRC Press, IGI Global, Auerbach. She is also co-author of two books published by NCERT of XI and XII and co-editor in 10 edited books. She has also organized 05 International conference sessions, 03 Faculty development Programmes, 01 workshop, 01 industrial meet in her experience. She delivered an expert talk, guest lecturers in International Conference and a member of reviewer/technical program committee in various International Conferences. Besides this, she associated with many International research organizations as Associate Editor/Guest Editor of Springer, Wiley and Elsevier books, and a reviewer for various International Journals.

Contributors

Mohammad Nasar Arbab
Department of Computer Science
and Engineering
Ambedkar Institute of Advanced
Communication Technologies &
Research
New Delhi, India

Ashirwad Barnwal
Department of Computer Science
and Engineering
Ambedkar Institute of Advanced
Communication Technologies
& Research
New Delhi, India

Renu Dalal
Department of Computer Science
and Engineering
Ambedkar Institute of Advanced
Communication Technologies
& Research
New Delhi, India

Sachin Dhawan
Department of Electronics
& Communication Engineering
Ambedkar Institute of Advanced
Communication Technologies
and Research
New Delhi, India

Deepti Dhingra
Department of Computer Science
and Engineering
Panipat Institute of Engineering
and Technology
Haryana, India

Avinash Dubey
Department of Computer Science
and Engineering
Ambedkar Institute of Advanced
Communication Technologies
& Research
New Delhi, India

Archit Garg
Department of Computer Science
and Engineering
Ambedkar Institute of Advanced
Communication Technologies
& Research
New Delhi, India

Anubhav Gautam
Department of Computer Science
and Engineering
Ambedkar Institute of Advanced
Communication Technologies
& Research
New Delhi, India

Nidhi Goel
Department of Electronics
& Communication Engineering
IGDTUW
New Delhi, India

Ashish Gupta
Department of Electronics
& Communication Engineering
North Eastern Regional Institute of
Science and Technology, Nirjuli
Itanagar, Arunachal Pradesh, India
and
ABES Engineering College
Ghaziabad, Uttar Pradesh

Dheeraj Gupta
Department of Computer Science
    and Engineering
Ambedkar Institute of Advanced
    Communication Technologies
    & Research
New Delhi, India

Rashmi Gupta
Department of Electronics
    & Communication Engineering
NSUT
Delhi, India

Manju Khari
Department of Computer Science
    and Engineering
Ambedkar Institute of Advanced
    Communication Technologies
    & Research
New Delhi, India

Rajesh Kumar
Department of Electronics
    and Communication
    Engineering
North Eastern Regional Institute
    of Science and Technology,
    Nirjuli
Itanagar, Arunachal Pradesh, India

Harshit Maheshwari
Department of Computer Science
    and Engineering
Ambedkar Institute of Advanced
    Communication Technologies
    & Research
New Delhi, India

Monika Mathur
Department of Electronics
    & Communication Engineering
Research Scholar, IGDTUW
Delhi, India

Bhavya Mehta
Department of Computer Science
    and Engineering
Ambedkar Institute of Advanced
    Communication Technologies
    & Research
New Delhi, India

Pooja Pandey
Department of Electronics
    & Communication Engineering
Research Scholar, IGDTUW
Delhi, India

Arun Kumar Rana
Department of Electronics
    & Communication Engineering
Panipat Institute of Engineering
    and Technology
Samalkha, Haryana, India
and
Department of Electronics
    & Communication Engineering
Maharishi Markandeshwar
    (Deemed to be University)
Mullana, Haryana, India

Sanjeev Kumar Saini
Department of Electronics
    & Communication Engineering
ABES Engineering College
Ghaziabad, Uttar Pradesh, India

Guru Gobind Singh
Indraprastha University
Dwarka, New Delhi, India

Arti Sharma
Department of Computer Science
    and Engineering
Ambedkar Institute of Advanced
    Communication Technologies
    & Research
New Delhi, India

Arush Sharma
Department of Computer Science
  and Engineering
Ambedkar Institute of Advanced
  Communication Technologies
  & Research
New Delhi, India

Sharad Sharma
Department of Electronics
  & Communication Engineering
Maharishi Markandeshwar
  (Deemed to be University)
Mullana, Haryana, India

Shubham Tayal
Department of ECE
SR University
Warangal, Telangana,
  India

Devvrat Tyagi
Department of Electronics
  & Communication
  Engineering
North Eastern Regional Institute
  of Science and Technology,
  Nirjuli
Itanagar, Arunachal Pradesh, India

1 Deep Learning-Based Computer Vision: Security, Application and Opportunities

Deepti Dhingra
Panipat Institute of Engineering and Technology

Sachin Dhawan
Ambedkar Institute of Advanced
Communication Technology & Research

Rashmi Gupta
NSUT East

CONTENTS

DOI: 10.1201/9781003138068-1

1.1   INTRODUCTION

Computer Vision, also referred to as CV, is defined as a branch of computer science that develops techniques through which computers can "see" and understand, interpret the digital images and videos. Our brain can identify the thing, process data and choose what to try to do, thus can do a difficult task during a blink of an eye. The aim is to form computers to be ready to do an equivalent. Hence, it is a field which will be called as a mixture of artificial intelligence (AI) and machine learning, which have learning algorithms and special methods to know what the pc sees.

1.1.1   TASKS IN CV

There are many tasks in CV which are utilized in world; for instance, once we look an image we will identify various objects therein, we will also find the situation and differentiate between background and object. Various subtasks involved in CV are as follows: Semantic Segmentation, Image Classification, Action Detection, Gesture Recognition, Object Tracking, Emotion Recognition, Object Localization, 3D Absolute Human Pose Estimation, Deep Fake Detection, Face Generation, 3D Object Detection, Facial Expression Recognition, Intelligent Surveillance, Human-Object Interaction Detection, Video Object Segmentation, Action Detection.

1.1.2   APPLICATIONS OF CV

CV is a field of science that creates computers or devices to interpret different objects a bit like human see. The computers got to be trained to detect objects and also some patterns a bit like you teach a child to spot the objects but the computers are more efficient because it takes little or no time to be trained. CV has applications in industries and sectors and they are as follows:

a. **Oil and Gas**: The oil and natural gas companies produce many barrels of oil and billions of cubic feet of gas a day except for this to happen; first, the geologists need to find a feasible location from where oil and gas are often extracted to seek out these locations they need to research thousands of various locations using images taken on the spot. Suppose if geologists had to research each image manually how long will it fancy find the simplest location? Maybe months or may be a year but thanks to the introduction of CV the amount of analyzing are often brought right down to a couple of days or may be a couple of hours. You only have got to feed the pictures taken to the pre-trained model and it will get the work done.

b. **Hiring Process**: Within the HR world, CV is changing how candidates get hired within the interview process. By using CV, machine learning, and data science, they are ready to quantify soft skills and conduct early candidate assessments to assist large companies shortlist the candidates.

c. **Video Surveillance**: The concept of video tagging is employed to tag videos with keywords supporting the objects that appear in each scene. Now imagine being that security company who is asking to seem for a suspect during a blue van among hours and hours of footage. You will just need to feed the video to the algorithm. With CV and visual perception, rummaging through videos has become a thing of the past.

d. **Construction**: For example the electrical towers or buildings, which require a point of maintenance to see for degrees of rust and other structural defects. Certainly, manually climbing up the tower to seem at every inch and corner would be extremely time-consuming, costly and dangerous. Flying a drone with wires round the electric tower does not sound particularly safe either. So how could you apply CV here? Imagine that if an individual on the bottom took high-resolution images from different angles. Then the pc vision specialist could create a custom classifier and use it to detect the issues and amount of rust or cracks present.

e. **Healthcare**: From the past few years, the healthcare industry has adopted many next-generation technologies that include AI and machine learning concept; one among them is CV which helps determine or diagnose disease in humans or any living creatures.

f. **Agriculture**: The agricultural farms have started using CV technologies in various forms like smart tractors, smart farming equipment, and drones to assist monitor and maintain the fields efficiently and simply. It also helps improve yield and therefore the quality of crops.

g. **Military**: CV is a crucial technology that helps military them to detect enemy troops and it also helps to target missile systems. It uses image sensors that deliver intelligent information about the battlefield. Autonomous vehicles and remote-controlled semi-automatic vehicles are also an important application of CV.

h. **Industry**: In manufacturing or production line, automatic inspections like identifying defective products on the assembly line, finding defects in packaging, problems in completed product can be done using CV. The technology is also applied to increase the efficiency of the assembly line.

i. **Automotive**: This is often one among the simplest samples of CV technologies, which may be a dream come true for humans. Self-driving AI-based cars analyzes data from a camera on the vehicle to find automatic lane, object detection, and recognize pedestrian, traffic signs and signals.

j. **Automated Lip Reading**: this is often one among the sensible implementations of CV to assist people with disabilities or who cannot speak; it reads the movement of lips and compares it to already known movements that were recorded and wont to create the model.

1.1.3 OBJECT DETECTION

From a digital image or a video, an object detection model can find any object from the input known set of objects and also provide its location and probability of its existence by forming bounding boxes on the objects. Means an object detection model identify and localized objects in an image or a video with its position. Object detection is used in many applications, for example: (i) human and computer interaction (HCI); (ii) object detection by any robot (e.g., service robots); (iii) image recognition in smart phone camera; (iv) face recognition and tracking used in (v) autonomous vehicles' uses for pedestrian detection.

An object detection model finds bounding boxes, one for every object it also finds, also as probabilities of its findings for every object. An object detection model may find too many bounding boxes of an object. Each box also has a confidence score that gives the chances of presence of the object in image. At step two we filter the boxes whose score falls below a particular threshold (also called non-maximum suppression). When we apply object detection model on an image it will give output as all the objects it detects with bounding boxes on each object and score on the box which give the percentage of its presence in image.

An object detection model is applied to find the presence of fruit in the provided image. The model is provided with multiple classes of fruits as input. Then the model is trained on large dataset of all images of those fruits. Then if we want to find the fruit detection in that model with labeling an apple, a banana, or a strawberry and data also tell that where each object appears within the image that is its location within image.

So we will model a system which we have trained to detect apples, bananas and straw berries. Once we pass it image, it will give output as follows:

1.1.3.1 Model Output

Class Score Location
Apple 0.92 [18, 21, 57, 63]
Banana 0.88 [100, 30, 180, 150]
Strawberry 0.87 [7, 82, 89, 163]
Banana 0.23 [42, 66, 57, 83]
Apple 0.11 [6, 42, 31, 58]

To understand these results, we have to check the score and the location for every detected object. The confidence score is any number between 0 and 1. If the number is close to 1 then it is accurately detecting. And if the score is close to 0 then it means that the chances of its occurrence in the image are very less and it is false prediction. A threshold value of confidence score can also be taken to discard false predictions means that these objects are not found in the image. For our example, we decide a confidence score of 0.5 means if probability of detection is above 50% then it is correct detection.

1.1.4 IMAGE CLASSIFICATION

The task of identifying what a picture represents is named image classification. A picture classification model is trained to acknowledge various classes of images; for instance, you will train a model to acknowledge photos representing differing types

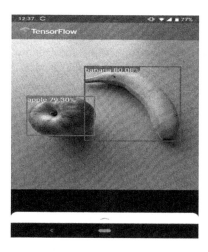

FIGURE 1.1 Example object detection [2].

of classes, animals, rabbits, and dogs. Figure 1.1 shows the sample of object detection in Tensor flow. Figure 1.2 show the sample of image classification using various objects.

1.1.5 IMAGE SEGMENTATION

In image segmentation picture is divided into regions. Classification is done on the basis of pixel. Picture is divided into several parts. It is used to separate foreground image from background. Pixels are grouped together or kept in same part if they are similar in shape size or color. Image segmentation has large number of applications; for example, it is used in medical diagnosis to separate tumor. In

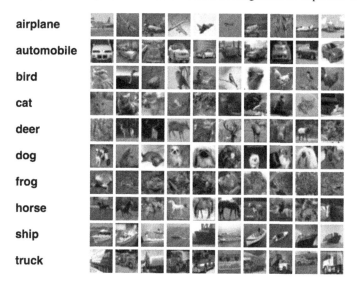

FIGURE 1.2 Image classifications [6].

autonomous vehicles it is used to separate objects with proper label from its background image; segmentation is of two types: (i) semantic segmentation, in which relation between segments is find out; (ii) instance segmentation, which finds different labels for various object instances in the picture. It is an extra feature to semantic segmentation.

1.1.6 DEEP LEARNING

Deep learning is a subset of machine learning method that makes computers to understand and behave same as humans. The technology behind self driving cars, which helps them to understand and acknowledge a stop sign, or to detect to a pedestrian in on road. All voice control features in electronic devices like phone, speaker, vision features of phone like camera, etc., are examples of deep learning. This technology has become a powerful tool and can be applied on all the sectors. It results in a great accuracy and speed in real-time problems. If we are provided with a class of objects then deep learning methods can be used to train a computer and identify any image from the given class of images, text or sound. Deep learning models are easy, fast and accurate, to do such kind of difficult AI tasks. These algorithms sometimes are more accurate than humans. These deep learning models are provided with a large set of labeled data and neural networks having many hidden layers in them are trained to do the tasks.

Deep learning models use neural network to train a data which gives good accuracy. In these models a large number of hidden layers are used to train the data. These architectures make the machine. Deep learning models are used for image recognition, object detection, image segmentation and many complex tasks of AI in various domains like health and self-driving car. In some tasks deep learning models have given more performance than humans like image classification.

1.1.7 SAMPLES OF DEEP LEARNING AT WORK

Deep learning models have applications in all the industries. Few of them are as follows:

Automated Driving: In autonomous vehicles deep learning is used to train the machine so that it can detect object, traffic signals, pedestrian and can also do image segmentation. Deep learning is a powerful tool which makes machine to learn.

Aerospace and Defense: Deep learning is applied on images captured from satellites to identify spot objects. It is used in security purpose and to locate safe and unsafe zone.

Medical Research: Deep learning methods are applied in medical science to detect cancer cells. A large set of data is provided to train the system which further can detect cancer.

Industrial Automation: In all the industries manufacturing, packaging and quality check all tasks in industries are done automatically and deep learning model are trained to do all the tasks like detection of defects, etc., accurately.

Electronics: Voice-based assistant, hearing and speech generation, translation, natural language understanding and many more tasks are performed by deep learning models.

1.1.8 CONVOLUTION NEURAL NETWORKS

Most deep learning methods are based on neural network architectures with a large number of hidden layers. In deep learning "deep" terms refers to a large number of hidden layers used in neural network. Simple neural networks have two to three hidden layers, while deep networks can have 150 or more layers; large labeled dataset is used to train deep learning models and automatically neural network extracts features from the images after training. No manual feature extraction is required in neural networks. Convolution neural network (CNN or ConvNet) is the most important deep learning model. Figure 1.3 shows Convolution neural network for object detection. Figure 1.4 shows convolution layer of an image. Figure 1.5 shows pooling layer.

An example to illustrate the concept of a convolution layer:

The below image clarifies the concept of a pooling layer (Average or Max pooling):

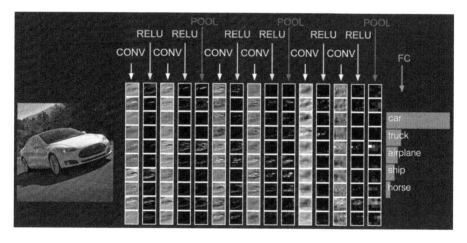

FIGURE 1.3 Convolution neural network [8].

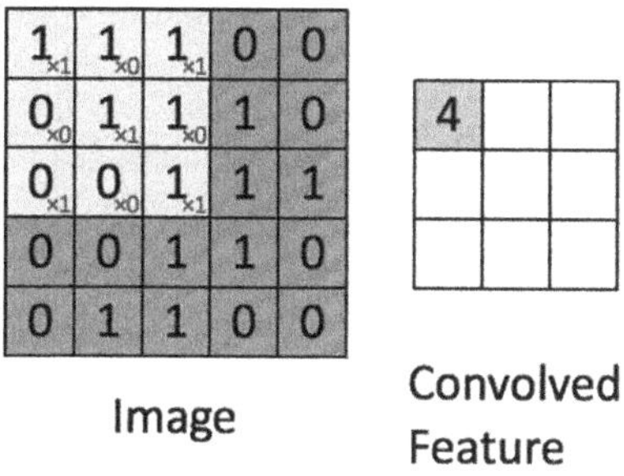

FIGURE 1.4 Convolution layer [8].

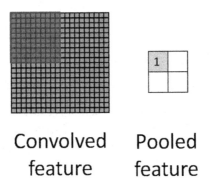

Convolved   Pooled
feature   feature

FIGURE 1.5   Pooling layer [9].

1.2   RESEARCH CHALLENGES OF CV

- Rapid recent progress in the field of CV has had a significant real-world impact, opening possibilities in domains such as transportation, entertainment and safety. While these are valuable and meaningful technological applications, CV has the potential to benefit people around the globe in many other ways, as well, in fields such as agriculture, disease prevention, infrastructure mapping, and more. These applications can lead the CV community to discover intellectually interesting and challenging new tasks and data sets and, by broadening the problem statements and their geographic diversity, help to further expose biases in the tasks and data sets.
- The world open global problems must be solved by present new techniques and solutions. The visual world is diverse, and technology must account for this diversity.
- Artificial intelligence must improve access to healthcare, accelerate economic development, reduce inequalities and other gains.
- Apply CV technology in support of efforts to address global poverty and inequality.

1.3   OBJECT DETECTION METHODS

There are two types of object detection methods: one-stage method and two-stage method. Figure 1.6 shows the classifications of methods of object detection.

1.3.1   YOLO (You Only Look Once)

In [1] object detection is taken as a single regression problem, it is done by converting pixels from the image into 0 bounding box coordinates and class probabilities. Using this system, you only look once (YOLO) an object can be detected in an image with its location. This is a single convolution network; it can find many bounding boxes and class probabilities at the same time in an image. Full image is trained on YOLO and the detection is done which will increase the performance. It is a fast method of image detection. It needs a complex structure to detect the object. When simple

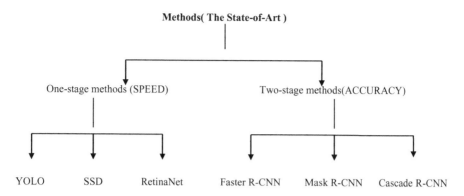

FIGURE 1.6   Methods of object detection [10].

neural network is implemented to detect then it was observed as a faster method it detects 45 frames per second. A fast version of YOLO detects 150 fps. Real video can be processed at a faster speed using YOLO. YOLO also gives twice the accuracy achieved by other methods. The *"You Only Look Once" is a* deep learning model made for fast object detection, framed by Joseph Redmon. The method involves a single deep CNN that divides the input into a grid of cells and each cell directly finds a bounding box and object classification. The output is a large number of bounding boxes that gives a final prediction.

1.3.2   MASK R-CNN

This algorithm is based on regions or parts of an image. It is implemented as regions with convolutional neural networks (R-CNN); R-CNN is a two-stage detection algorithm. The first step finds regions in an image and second stage will classify the object in each region.

Following are the features of RCNN:

- Identify regions in an image called as region proposals
- Find CNN features from the region proposals
- Classify the objects using the extracted features

There are three types of R-CNN. Each type is focused on optimization, speed or accuracy of the detection.

Comparison of YOLO and R-CNN

- Fastest method of object detection is Fast YOLO.
- YOLO gives mAP to 63.4%, thus a good performance for real-time objects.
- Faster R-CNN is 2.5 times slower than YOLO and it also gives lesser accuracy [2].

Object instance segmentation is presented in [2] which is a simple and good method in terms of performance. In this approach we efficiently detect objects in an image and divide it into different parts or segments like foreground and background.

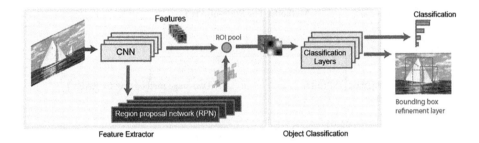

FIGURE 1.7 RCNN [11].

The method, called as Mask R-CNN, is better and it extends Faster R-CNN as it predicts object mask also. It also gives better speed than faster R-CNN as running time increases by 5 fps. It can be used to estimate human poses. Figure 1.7 shows the block diagram of Recurrent neural network for object detection.

1.3.3 SDD (Single Shot Multi-Box Detector)

The SSD approach proposed in [3] is also a convolution network which is feed forward. This method gives a fixed-size collection of bounding boxes and scores that tell the presence of object class in those bounding boxes as output. After this some threshold is decided to take final detections. The starting network layers are based on a simple image classification models called as base network. After that an auxiliary structure is added to the network to get detections which must have features like multi-scale maps for detection.

1.3.4 Retina Net

Retina Net is an object detection method with backbone network and two task-specific sub networks presented in [4]. Feature map of an input image is taken by the backbone network which is convolutional network. The first convolutional subnet has responsibility to perform object classification on the backbone's output; the second subnet convolutional has to perform bounding box regression. The two sub-networks make a simple design for detection which is proposed for one stage. There are many choices to select parameters for these sub-networks.

Feature Pyramid Network [5] is used by one-stage Retina Net. Backbone network is feed forward ResNet architecture [6]. To get multi-scale convolution feature pyramid backbone Retina Net has two sub-networks, one for classifying anchor boxes. One subnet is used for regressing and the network design is simple so that accuracy can be achieved.

1.3.5 Faster R-CNN

A faster R-CNN network presented in [7] takes complete image and a set of object proposals as input. The network in first step uses many convolution (conv) and max pooling layers to give a convolution feature map of an image. In the next step pooling layer of the model will extract a fixed-length feature vector from each proposal. Then, each feature vector is given as an input into a sequence of fully connected (fc) layers that in the end is connected two output layers. Fast R-CNN is a faster method as compared to R-CNN and SPPnet. It has shown state-of-the-art detection results. Sparse object proposals prove that they improve detector quality. It was too costly as it takes a lot of time, but it becomes practical with Fast R-CNN.

1.3.6 CASCADE R-CNN

To define positives and negatives in object detection, an intersection over union (IoU) threshold is needed to define. An object detector, trained with low IoU threshold, e.g., 0.5, gives noisy detections, and if we increase IoU thresholds then the performance degrades. Two responsible factors for the same are as follows: (i) over-fitting is caused due to loss of positive samples; (ii) inference-time mismatch. The cascade R-CNN is a multi-stage object detection architecture proposed in [8] to solve these problems. This architecture has a sequence of detectors to avoid false positives and the training of detectors is done by increasing the IoU threshold. The detectors are trained stage by stage, so that at each level of detection the quality get improves. Over-fitting problem is solved by resampling. It is a better method of object detection as it improves the detection process sequentially.

1.4 IMAGE CLASSIFICATION MODELS

1.4.1 ALEXNET

The architecture contains eight layers: five convolutional layers and three fully connected layers. But this is not the only main point. Its main features are as follows:

- **ReLU Nonlinearity**. AlexNet uses Rectified Linear Units (ReLU) function; at that time it was really a unique function to apply. Usage of ReLU's benefited at that time because at that time tanh function was used in general and using ReLU gave more accuracy and speed.
- **Multiple GPUs**. At that time GPU was running only with 3 GB of memory. As the data set had 1.2 million images, AlexNet was trained by multi-GPU by putting half neurons on one GPU and half on another GPU. Thus training time also reduces.
- **Overlapping Pooling**. Originally CNNs "pool" outputs of neighboring groups of neurons with no overlapping. But in AlexNet authors overlapped the output and they observed that error reduces by 0.5% and they found that models with overlapping pooling are not easy.
- **The Overfitting Problem**. AlexNet was having 60 million parameters, which gives a big issue like overfitting. Two methods were used to reduce overfitting.
- **Data Augmentation**. Data augmentation was used by authors in which they did label preserving transformation. They did image translations and horizontal reflections which helped to increase the dataset as data get varied by

doing so. As a result training set gets increased by a factor of 2,048. To change the intensity they performed Principal Component Analysis on the RGB channels. This reduces top 1 error rate by more than 1%.

- **Dropout**. Neurons in the model are assigned probability (e.g., 50%) in advance. Neurons are turning off as this technique consists of "turning off" neurons. So that in training random neurons can be used which give robust features and dropout also increases the training time.
- In [9] AlexNet network model combined with SSD model is used for the image recognition in tennis court. AlexNet is trained to test if the tennis ball is present or not in the court means that it will perform object detection task and SSD model is used to give exact location of the ball.

1.4.2 VGGNet

The VGG network architecture was developed by Simonyan and Zisserman. In this model an image of 224×224 RGB is given as input, then subtract mean RGR value from each pixel as a pre-processing step. During training the model, presented in [10], the image goes through a stack of convolution (conv.) layers, and there we use filters of $3 \times 3.1 \times 1$ convolution filters are also used in training. Input channels are also linearly transformed. The convolution stride is taken as 1 pixel; the spatial padding of conv. layer is taken as 1 pixel for 3×3 conv. layers. Some convolution layers are followed by five max-pooling layers. Max-pooling is done over a 2×2 pixel window, with stride 2.

Steps involved in training the model are as follows:

Step1: Load the data (get all working directories ready, input the images, resize if required)
Step2: Configure the model (transform the data; build the model, set up the parameters)
Step3: Train the model (train the model)
Step4: Evaluation and prediction (get output predictions and class labels)

1.4.3 ResNet

ResNet is different than AlextNet and VGGNet as these are sequential network models. ResNet has a special architecture. It has micro components. These are also called "network-in-network architectures." Building blocks are micro in these types of networks and collection of micro-components gives rise to large or macro networks.

ResNet introduced in [11] is a deep neural network based on residual networks. It has a residual module. It gives accuracy due to modifications in residual network. ResNet uses large number of hidden layers as compared to VGG16 and VGG19 [12].

ResNet architecture is used in remanufacturing inspection based on deep learning [13]. These object detection and classification methods are used to find defects in products and select them to remanufacture.

1.4.4 SQUEEZNET

In [14] the method SqueezNet is defined which uses less number of parameters and CNN is trained to give good accuracy. There are three ways to design CNN for better accuracy:

- Strategy 1: Replace 3×3 filters with 1×1 filters. Thus less number of parameters will be used; hence accuracy will get increased.
- Strategy 2: Decrease the number of input channels to 3×3 filters. If we decrease the number of input channels then the total number of parameters gets reduced; hence we are squeezing the network. Hence the accuracy increases.
- Strategy 3: We wish to have a large activation map so down-sampling should be done late at the network. We have to control the height and width of activation maps; this can be controlled by (i) the size of the input data (e.g., 256×256 images) and (ii) choose the layers in which we have to do down-sampling.

1.4.5 GOOGLENET

GoogleNet architecture proposed in [15] is designed in such a way that the depth, that is the number of layers, is increased in such a way that resources are efficiently used and computation cost is less. Multi-scale processing is used to get better quality. GoogleNet is a deep network with 22 layers used for classification and detection. 1×1 convolution and global average pooling are used in architecture.

- **1×1 convolution:** 1×1 convolution is used in the architecture. These convolutions decrease the number of parameters. Thus by decreasing the parameters, the number of layers increases.
- **Global average pooling:** In this architecture in the last pooling layer is used which reduces the parameters which need to be trained. Hence it increases the efficiency.

1.5 RESEARCH GAPS

- As per the real object detection and classification it required that algorithms should not only accurately classify and localize objects from images and videos but also these algorithms must be fast.
- Image classification problems have class imbalance, which means in image segmentation tasks only class of objects is identified and the rest of the picture is treated as background.
- Object detection datasets have only few, while image classification datasets have more than 100,000 classes. From real video sources image classification data are generated with classes. But gathering accurately bounding boxes with proper labels from live streaming remains a great work [18].
- Increasing the number of hidden layers in the convolution training is not optimal and efficient on GPU [16].

- If we want to optimize accuracy then computational budget [17] gets compromised.
- Some methods may be discovered that can detect dense boxes to perform as well as sparse proposals.

1.6   CONCLUSION AND FUTURE WORK

- CV is a challenging research area and is applicable in all the domains. With powerful deep learning models it has gained a great attention and solved complex problems of AI in a beautiful manner. In this report, I conduct a comprehensive survey on CV techniques and tasks. The basis of deep learning model is CNN, with the help of which machines can perform all the tasks with much ease, speed and accuracy as the human does. In some cases machines are more powerful than humans [19]. All the network architectures for object detection and image classification are studied in detail.
- Work can be extend in multi-task optimization in some applications like instance segmentation [5], multi-object tracking [20], and multi-person pose estimation.
- Performance of all the models quickly degrades when the quality of image is low. We need to have a model which can detect from partially labeled data.
- In the future, efficient DNNs (deep neural network) are needed for real-time and embedded applications. For example, state-of-the-art methods of object detection methods using deep learning can be applied to extract frames from live videos and can be applied for video surveillance and further crime predictions and avoiding by alarming any near future suspicious activity and many more.

REFERENCES

1. J. S. D. R. G. A. F. Redmon, "(YOLO) you only look once," *Cvpr*, 2016, doi: 10.1109/CVPR.2016.91.
2. K. He, G. Gkioxari, P. Dollár, and R. Girshick, "Mask R-CNN," *IEEE Trans. Pattern Anal. Mach. Intell.*, vol. 42, no. 2, pp. 386–397, 2020, doi: 10.1109/TPAMI.2018.2844175.
3. W. Liu *et al.*, "SSD: Single shot multibox detector," *Lect. Notes Comput. Sci. (including Subser. Lect. Notes Artif. Intell. Lect. Notes Bioinformatics)*, vol. 9905 LNCS, pp. 21–37, 2016, doi: 10.1007/978-3-319-46448-0_2.
4. Y. Li and F. Ren, "Light-weight retinaNet for object detection." Accessed: Dec. 4, 2020. [Online]. Available: https://github.com/PSCLab-ASU/LW-RetinaNet.
5. T.-Y. Lin, P. Dollár, R. Girshick, K. He, B. Hariharan, and S. Belongie, "Feature pyramid networks for object detection." *2017 IEEE Conference on Computer Vision and Pattern Recognition (CVPR)*, pp. 936–944, 2017, doi: 10.1109/CVPR.2017.106.
6. K. He, X. Zhang, S. Ren, and J. Sun, "Deep residual learning for image recognition," *Proc. IEEE Comput. Soc. Conf. Comput. Vis. Pattern Recognit.*, vol. 2016, pp. 770–778, 2016, doi: 10.1109/CVPR.2016.90.
7. R. Girshick, "Fast R-CNN," *Proc. IEEE Int. Conf. Comput. Vis.*, vol. 2015 Inter, pp. 1440–1448, 2015, doi: 10.1109/ICCV.2015.169.
8. Z. Cai and N. Vasconcelos, "Cascade R-CNN: Delving into high quality object detection." Accessed: Dec. 5, 2020. [Online]. Available: https://github.com/zhaoweicai/cascade-rcnn.

9. S. Gu, L. Ding, Y. Yang, and X. Chen, "A new deep learning method based on AlexNet model and SSD model for tennis ball recognition," *2017 IEEE 10th Int. Work. Comput. Intell. Appl. IWCIA 2017-Proc.*, vol. 2017, pp. 159–164, 2017, doi: 10.1109/IWCIA. 2017.8203578.

10. K. Simonyan and A. Zisserman, "Very deep convolutional networks for large-scale image recognition," *3rd Int. Conf. Learn. Represent. ICLR 2015-Conf. Track Proc.*, pp. 1–14, 2015.

11. K. He, X. Zhang, S. Ren, and J. Sun, "Deep residual learning for image recognition," *Proceedings of the IEEE Computer Society Conference on Computer Vision and Pattern Recognition*, vol. 2016-December, pp. 770–778, Dec. 2016, doi: 10.1109/ CVPR.2016.90.

12. K. He, X. Zhang, S. Ren, and J. Sun, "Identity mappings in deep residual networks," *Lect. Notes Comput. Sci. (including Subser. Lect. Notes Artif. Intell. Lect. Notes Bioinformatics)*, vol. 9908 LNCS, pp. 630–645, Mar. 2016.

13. C. Nwankpa, S. Eze, W. Ijomah, A. Gachagan, and S. Marshall, "Achieving remanufacturing inspection using deep learning," *J. Remanufacturing*, 2020, doi: 10.1007/ s13243-020-00093-9.

14. F. N. Iandola, S. Han, M. W. Moskewicz, K. Ashraf, W. J. Dally, and K. Keutzer, "SqueezeNet: AlexNet-level accuracy with 50x fewer parameters and <0.5MB model size," Feb. 2016. Accessed: Dec. 5, 2020. [Online]. Available: http://arxiv.org/abs/ 1602.07360.

15. C. Szegedy *et al.*, "Going deeper with convolutions," *Proc. IEEE Comput. Soc. Conf. Comput. Vis. Pattern Recognit.*, vol. 7–12, pp. 1–9, 2015, doi: 10.1109/CVPR.2015.7298594.

16. H. Zhang *et al.*, "ResNeSt: Split-attention networks," Computer Vision and Pattern Recognition *arXiv*, 2020.

17. X. Feng, Y. Jiang, X. Yang, M. Du, and X. Li, "Computer vision algorithms and hardware implementations: A survey," *Integration*, vol. 69. pp. 309–320, 2019, doi: 10.1016/j. vlsi.2019.07.005.

18. S. Ren, K. He, R. Girshick, and J. Sun, "Faster R-CNN: Towards Real-Time Object Detection with Region Proposal Networks," 2015. Accessed: Dec. 5, 2020. [Online]. Available: http://image-net.org/challenges/LSVRC/2015/results.

19. Z. Q. Zhao, P. Zheng, S. T. Xu, and X. Wu, "Object detection with deep learning: A review," *IEEE Trans. Neural Netw. Learn. Syst.*, vol. 30, no. 11, pp. 3212–3232, 2018.

20. S. Tang, M. Andriluka, and B. Schiele, "Detection and tracking of occluded people," *Int. J. Comput. Vis.*, vol. 110, no. 1, pp. 58–69, Oct. 2014, doi: 10.1007/s11263-013-0664-6.

2 Recognition of Foggy Image for Surveillance Application

Pooja Pandey
IGDTUW

Rashmi Gupta
NSUT Eastrashmi.gupta@nsut.ac.in

Nidhi Goel
IGDTUW

CONTENTS

2.1 INTRODUCTION

In foggy weather condition, size of atmospheric particles increases almost five times as compared to clear weather condition [1]. Due to this, light reflected from the image during image capturing got attenuated and scattered in its path. Attenuation and scattering of light cause loss of energy, and thus, quality of image in foggy weather degrades.

For restoration of foggy image, it is important to understand the physical model of foggy image formation. Actual model of image formation during poor weather condition is quite complex and it depends on various parameters like nature of atmospheric particles in terms of size, orientation and distribution. Also, it depends on the nature of incident light in terms of wavelength, direction and polarization. Physical model of foggy image formation is highly complex if all these parameters are considered. Taking this into consideration, Narsimhan et al. [2,3] has proposed simplified

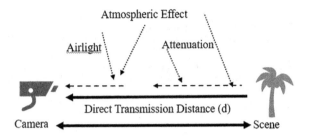

FIGURE 2.1   Atmospheric scattering model for foggy image.

physical imaging model based on atmospheric scattering phenomenon as shown in Figure 2.1. Mathematically, this physical model is represented using Equation (2.1).

$$I_f(a, b) = I(a, b)\Gamma(a, b) + A(1 - \Gamma(a, b)) \tag{2.1}$$

where $I_f(a, b)$ is foggy image; $I(a, b)$ is fog free image or clear image; $\Gamma(a, b)$ is transmission map; A is atmospheric light. Equation (2.1) is the general equation which is used in many image processing and computer vision algorithm. In Equation (2.1), two parameters Γ and A are unknown. Once these parameters are estimated, fog-free image can be calculated using Equation (2.2).

$$I(a, b) = \frac{I_f(a, b) - A}{\Gamma(a, b)} + A \tag{2.2}$$

In the proposed method, foggy image is pre-processed using white balancing approach. At the intermediate stage, guided filter is used to extract finer details of image and to give smooth output. To further enhance the contrast of image, partially overlapped sub-block histogram equalization [4] has been used.

The rest of the chapter is organized as follows. In Section 2.2, state-of-the-art methods are briefly discussed. Section 2.3 gives idea about the different technique which has been used in the proposed work. In Section 2.4, details of proposed work are explained in a step-by-step manner. Results of the proposed work and its quantitative and qualitative comparison with existing methods are shown in Section 2.5. Finally, in Section 2.6, the proposed method is concluded.

2.2   LITERATURE REVIEW

Fog removal has always been a challenging task for many researchers due to its dependency on various unknown parameters like scene depth and atmospheric lights. Researchers in earlier defogging techniques use more than one image to extract information required for restoring the foggy image. Polarization-based techniques are used in multiple image defogging methods. Two or more images with different polarization are considered under same weather conditions or different weather conditions [5,6]. In some defogging methods, scene depth information is used to restore image [7,8]. The main disadvantage of scene depth-based method and multiple image-based method is that in many real application information are not available as required by these algorithms. Due to these limitations, in recent years researcher use single foggy image to extract all information required for defogging of image.

In single image defogging methods, some methods are based on enhancement concept. Contrast and color of the foggy images are enhanced using different techniques like histogram [9], contrast [10], fusion [11,12] and Retinex [13,14]. Images obtained after enhancement based defogging methods are visually appealing but it looks artificial since parameters like contrast and color are enhanced without taking into consideration the actual scene information. There are some prior-based defogging methods which are based on some observation or assumptions to solve the ill-posed problem. In [15], defogging method is based on the assumption that fog-free images have higher contrast as compared to foggy image. And on the basis of this observation, contrast of foggy image is maximized to get fog-free image. But output shows oversaturation of image, and thus, defog image loses natural instinct. In [16], assumption has been made that there is no local relation between transmission and surface shading. But this assumption fails in case of dense fog.

Dark channel prior (DCP) proposed by He et al. [17] is motivated by technique dark-object subtraction. It is based on the observation that non-sky images have some pixels whose intensities are very low (approximately zero) in at least one of the color channels. In DCP, soft matting technique is used to refine transmission parameter which has high computational complexity. Due to this reason, overall execution time of DCP is high and not suited for real-time application. Other priors such as color attenuation prior (CAP) [18] and haze line prior (HLP) [19] are also proposed to remove fog from the image. CAP is based on the assumption that brightness of foggy image is high as compared to fig free image while saturation of foggy image is low as compared to fog-free image. But the different parameters calculated in this prior are based on learning model which requires large number of data sets. Some other defogging methods are based on inhomogeneous scattering model in which scattering coefficients are not considered as constant. In [20], defogging method is based on atmospheric point spread function and it considers scattering coefficient as variable not constant unlike many conventional methods. In [21], Ju et al.'s proposed method improved atmospheric scattering model based on multiple scattering coefficients. This model overcomes the problem of classical atmospheric scattering model.

In [22], gamma correction method is used to upgrade conventional atmospheric scattering model and a new model, gamma correction-based dehazing model, has been proposed. But the disadvantage of these methods [20–22] is high complexity. Also, models proposed do not give satisfactory result in real-time application. In recent years, machine learning-based defogging methods have also been proposed by many researchers. Tang et al. [23] and Kim et al. [24] proposed defogging methods based on Random Forest model. To train regression model for calculating different parameters required in image defogging, Random Forest is used. In [25], convolution neural network (CNN) has been used to learn the mapping relation between foggy image and transmission map. In [26], multi-scale deep neural network is used to estimate the transmission map of foggy image. In [27], both atmospheric light and transmission map are estimated using cascaded CNN.

2.3 PROPOSED METHOD

The block diagram of the proposed method is shown in Figure 2.2. Different steps which are involved in proposed work can be enumerated as:

Step 1. Input is foggy image

Step 2. White balancing method is used as a pre-processing technique. The main intent of white balance technique as a pre-processor is to find illuminant color $t(\lambda)$ in the RGB channel. Atmospheric light causes color shift and chromatic cast in foggy image. This chromatic cast can be minimized using white balancing technique. There is different white balancing approach which has been proposed by different researchers [28–30]. In the proposed method, shades-of-gray based on color constancy technique [29] has been used for white balancing of foggy image. Any image p in terms of Lambertian surface can be represented as:

$$p(x) = \int_0^\omega l(\lambda)m(, x)q(\lambda)d(\lambda) \tag{2.3}$$

where $l(\lambda)$ is radiance of light source, $m(\lambda, x)$ is surface reflectance, $q(\lambda)$ denotes sensors sensitivity and ω is spectrum of visible light. The illuminant color t can be estimated as:

$$t = (R_t, G_t, B_t) = \int_0^\omega l(\lambda)q(\lambda)d(\lambda) \tag{2.4}$$

Average reflectance of the scene is generally grey as per *Grey-World* theory [29]. Mathematically it can be defined as:

$$\frac{\int m(\lambda, x)d(x)}{\int d(x)} = 0.5 \tag{2.5}$$

Putting Equations (2.5) in (2.3), it can be written as:

$$\frac{\int p(x)d(x)}{\int d(x)} = 0.5 \int_0^\omega l(\lambda)q(\lambda)d(\lambda) \tag{2.6}$$

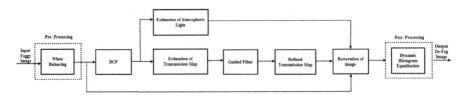

FIGURE 2.2   Block diagram of proposed method.

According to *shades-of-gray*, Minkowski norm is used to define white balance of image [29]. The average color of the entire image when raised to the power n is gray [31], and thus, it can be stated as:

$$\left[\frac{\int p^n dx}{\int dx}\right]^{\frac{1}{n}} = 0.5 \ t = 0.5(R_t, G_t, B_t) \tag{2.7}$$

Intensity of image depends on the value of *n*. According to *shades-of-gray* assumptions, range of *n* is from 1 to infinity. For $n = 1$, each component of the image gives uniform contribution.

Step 3. Parameter transmission map, required for image restoration, is calculated using DCP [17]. Dark channel of image '*I*' can be calculated as [17]:

$$I^{dark}(x) = \min_{y \varepsilon \Omega(a)}\left(\min_{c\varepsilon\{r,\,g,\,b\}} I^c(x)\right) \tag{2.8}$$

where I^c represents different color channel of image I and $\Omega(a)$ is a patch of size 15×15. According to DCP, dark channel of fog free image is:

$$I^{dark} \rightarrow 0 \tag{2.9}$$

Using Equations (2.1) and (2.9), transmission parameter is calculated as:

$$\Gamma(x)^{dark} = 1 - \min_{y \varepsilon \Omega(a)}\left(\min_{c\varepsilon\{r,\,g,\,b\}} \frac{I_f^c(x)}{A^c}\right) \tag{2.10}$$

Step 4. Transmission map calculated in step 3 is not smooth and contains some halos and artifacts. To remove this artifact, further refinement of transmission map is done using guided filter. In guided filter detail layers are computed and then combined for better edge preservation. For any guidance image I_G and target image I_T, the filtering output I_O for pixel position '*i*' and '*j*' can be written as:

$$I_{Oi} = \sum_{j} W_{ij}(I_G)I_{Tj} \tag{2.11}$$

Kernel of the filter 'W_{ij}' depends only on I_G and independent of I_T. There is a linear relation between input image and target image. In terms of filter's coefficient, output of the guided filter can be expressed as:

$$I_{Oi} = a_k I_{Gi} + b_k, \ i \in w_k, \tag{2.12}$$

where a_k and b_k are filter's coefficient and it is considered constant in window 'w_k'. Local linear relationship shows that output image 'I_O' has an edge only if guidance image 'I_G' has an edge. To determine filter coefficient, cost function which is minimized in the window is given as:

$$C(a_k, b_k) = \sum_{i \in w_k} \left((a_k I_{Gi} + b_k - I_{Ti})^2 + \omega a_k^2 \right) \qquad (2.13)$$

'ω' is regularization parameter and it prevents filter coefficients from very large value.

Linear regression method is used to find solution of cost function mentioned in Equation (2.8) and it is calculated as:

$$a_k = \frac{\frac{1}{|\omega|} \sum_{i \in w_k} I_{Gi} I_{Ti} - \mu_k \overline{I_T}_k}{\sigma_k^2 + \varepsilon} \qquad (2.14)$$

$$b_k = \overline{I_T}_k - a_k \mu_k \qquad (2.15)$$

where μ_k is mean and σ_k^2 is variance of guided image 'I_G' in w_k, $|w|$ is total pixels number in considered window, and

$$\overline{I_T}_k = \frac{1}{|\omega|} \sum_{i \in w_k} I_{Ti} \qquad (2.16)$$

Local window is used to cover entire image and after computing filter's coefficient (a_k and b_k) for each window 'w_k', output of the guided filter is obtained as:

$$I_{Oi} = \frac{1}{|\omega|} \sum_{k: i \in w_k} a_k I_{Gi} + b_k \qquad (2.17)$$

Thus, transmission map obtained in step 3 is refined using guided filter. Refined transmission map (Γ_{refined}) is further used for restoration of foggy image.

Step 5. Estimation of atmospheric light is also based on dark channel. Local region is selected in dark channel of image which contains top 0.1% brightest pixels, and then brightest pixel is selected from the original image corresponding to this local region. This brightest pixel is considered as atmospheric light A.

Step 6. Proposed method is based on restoration method and final restored image is given as:

$$I(a, b) = \frac{I_f(a, b) - A}{\Gamma_{\text{refined}}(a, b)} + A \qquad (2.18)$$

Step 7. Foggy image restored using Equation (2.18) are low in contrast. Low contrast images are visually not appealing and thus, image needs further processing. In proposed work, POSHE has been used to improve contrast of the image [4]. Main advantage of POSHE over other histogram equalization is that it enhances image contrast and also minimizes blocking artifacts. POSHE overcomes the

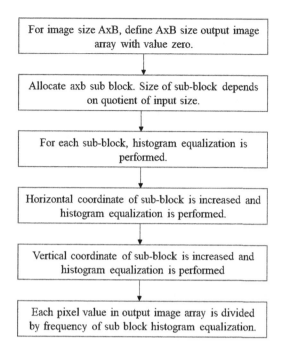

FIGURE 2.3   Different steps of POSHE.

limitation of both local histogram equalization (LHE) and global histogram equalization (GHE). Contrast enhancement achieved by LHE is more effective as compared to GHE but at the cost of high computational complexity. GHE has low computational complexity but at the cost of relatively low contrast enhancement. In contrary, POSHE ensures both high contrast enhancement and low computational complexity.

In POSHE, low pass filter (LPF) shaped mask is used. It is achieved by moving sub-blocks in a partially overlapped manner. Histogram equalization is performed on these sub blocks. In Figure 2.3 different steps involved in POSHE are shown.

2.4   RESULT ANALYSIS

Proposed algorithm is simulated on MATLAB software for various standard databases such as Haze-RD, FRIDA. Some images are taken from popular Google sources. The result of the proposed method is compared both qualitatively and quantitatively with state-of-the-art method.

2.4.1   QUALITATIVE ANALYSIS

Figures 2.4–2.9 show intermediate and final results of the proposed algorithm on six different data sets. In all these figures, (a) shows input foggy image, (b) shows transmission map using DCP, (c) shows refined transmission map using guided filter,

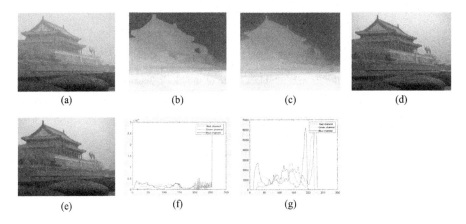

FIGURE 2.4   Output of proposed method on 1st foggy image.

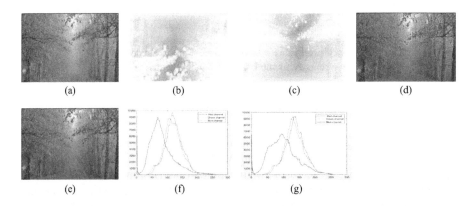

FIGURE 2.5   Output of proposed method on 2nd foggy image.

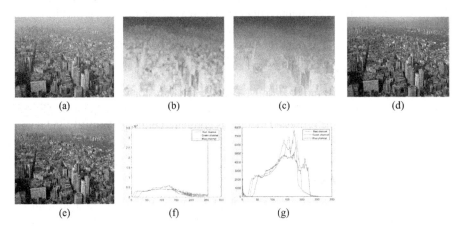

FIGURE 2.6   Output of proposed method on 3rd foggy image.

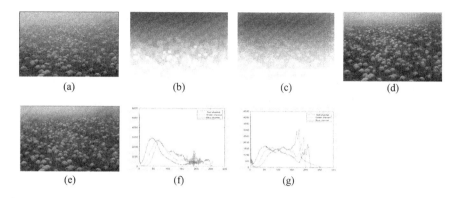

FIGURE 2.7 Output of proposed method on 4th foggy image.

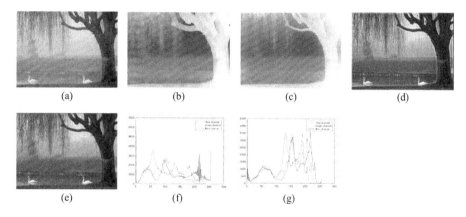

FIGURE 2.8 Output of proposed method on 5th foggy image.

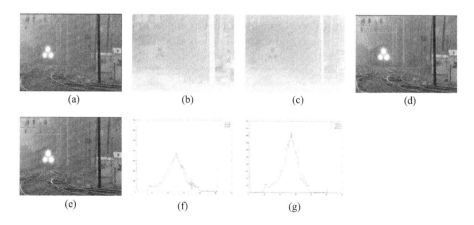

FIGURE 2.9 Output of proposed method on 6th foggy image.

(d) shows defog image using DCP, (e) shows output of proposed algorithm, (f) shows histogram of defog image using DCP and (g) shows histogram of defog image using proposed algorithm.

Defog output of proposed method shows that quality output image is enhanced and visually more appealing as compared to input image.

2.4.2　QUANTITATIVE ANALYSIS

To compare the efficiency of proposed algorithm different parameters such as: VM (visibility metric), CG (contrast gain), SSIM (structural similarity index measure) and time complexity are calculated and compared with different fog removal methods.

Visibility Metric (VM): This parameter is used in different image processing area to measure the visual enhancement performance. Mathematically this parameter can be calculated as:

$$VM = \frac{\mu_r - \mu_i}{\sigma_r} \qquad (2.19)$$

where μ_r and μ_i are mean of reference image and input image; σ_r is variance of reference image. Table 2.1 compares VM of the proposed method with state-of-the-art method.

Structural Similarity Index Measure (SSIM): It is a parameter which shows loss of information during transmission of signal from one place to other. It is full reference metric, and for its calculation, foggy image is considered as reference image. For calculating SSIM, three important factors are compared with reference image: luminance, contrast and structure. Small value of SSIM means better quality of output image because smaller value indicates that output image is less similar with the given input foggy image and that is the requirement of defogging algorithm. Table 2.2 compares SSIM of the proposed method with other defogging methods.

Contrast Gain (CG): Contrast gain value depicts the visual quality of any RGB image. Better contrast means image is visually more appealing as compared to low contrast image. So, CG parameter is very important parameter and it is used for quantitative analysis in different image processing algorithm. For calculating CG value, contrast of input image and output image is calculated and then compared.

TABLE 2.1
VM of Different Methods

Images	DCP [31]	BCP [13]	CAP [4]	HLP [5]	Proposed Method
FI1	58.0972	50.1094	71.0911	70.0008	65.7895
FI2	45.7813	42.1202	66.0071	54.0911	47.1875
FI3	72.2807	55.8111	81.0015	82.0076	83.3333
FI4	45.7350	40.0631	67.1200	63.0655	61.3430
FI5	35.5717	34.0822	64.0222	64.0110	44.1016
FI6	35.5717	35.0458	54.0661	56.2014	39.0923

TABLE 2.2
SSIM of Different Methods

Images	DCP [31]	BCP [13]	CAP [4]	HLP [5]	Proposed Method
FI1	0.6431	0.5805	0.7885	0.7698	0.7176
FI2	0.8761	0.7643	0.9805	0.9655	0.9231
FI3	0.7612	0.7505	0.8571	0.8218	0.7989
FI4	0.7861	0.7008	0.8733	0.8805	0.8015
FI5	0.8009	0.7621	0.8215	0.8109	0.8015
FI6	0.9231	0.8761	0.9654	0.9505	0.9433

TABLE 2.3
CG of Different Methods

Images	DCP [31]	BCP [13]	CAP [4]	HLP [5]	Proposed Method
FI1	0.0970	0.0902	0.1342	0.1176	0.1044
FI2	0.0388	0.0367	0.2009	0.1876	0.0379
FI3	0.2100	0.2008	0.2167	0.2166	0.2109
FI4	0.1285	0.1239	0.2109	0.1984	0.1430
FI5	0.0486	0.0457	0.1782	0.1865	0.1011
FI6	0.1350	0.1087	0.2059	0.2003	0.1905

TABLE 2.4
Time Complexity (in sec) of Different Methods

Images	DCP [31]	BCP [13]	CAP [4]	HLP [5]	Proposed Method
FI1	12.3427	16.9665	7.0088	8.5543	12.3326
FI2	22.7611	32.0094	14.3211	10.6653	18.0439
FI3	44.6681	55.2451	20.3377	15.0554	24.6319
FI4	18.0052	23.5004	10.7002	11.0333	20.8553
FI5	42.6112	51.0052	12.9911	9.0665	20.7741
FI6	33.9901	48.3112	9.0476	5.4040	25.6312

Table 2.3 shows CG of the proposed method and other methods. Algorithm is considered better if it has high CG. For an image of size $M \times N$, CG can be calculated as:

$$CG = \text{Mean contrast of output image} - \text{Mean contrast of input image} \quad (2.20)$$

$$\text{Mean Contrast of Image} = \frac{\sum_{a=0}^{M-1} \sum_{b=0}^{N-1} C(a, b)}{MN} \quad (2.21)$$

Time complexity of the proposed method is also compared with some of the existing methods for better analysis. Table 2.4 shows the processing time required in proposed algorithm and some of the existing defogging methods.

2.5 CONCLUSION

In foggy weather condition, object detection and tracking are difficult and it poses a threat to many security applications. To overcome such situations, improvement in foggy image quality is required. In the proposed method, different features like edge, texture, finer details of foggy image have been enhanced. Enhanced output image is compared with existing defogging methods. Quantitative and qualitative results show that the proposed method outperforms many existing methods and gives high-quality defog image. Object detection can be done easily from the defog image which is highly required in many security applications.

Time complexity of proposed method is less as compared to DCP and BCP as indicated by Table 2.4. But CAP and HLP methods requires less time for processing as compared to proposed method. High time of the proposed method is due to pre-processing and post-processing of image. In future work, this drawback of proposed method will be taken into consideration.

REFERENCES

1. G. M. Hidy and M. Kerker, "Aerosols and atmospheric chemistry: The Kendall award symposium honoring Milton Kerker," *Proceedings of the American Chemical Society, Los Angeles, California, March 28–April 2, 1971*. Academic Press. 1972.
2. S. G. Narasimhan and S. K. Nayar, "Removing weather effects from monochrome images," *Proc. IEEE Conf. Comput. Vis. Pattern Recognit. (CVPR)*, vol. 2, pp. 186–193, Dec. 2001.
3. S. G. Narasimhan and S. K. Nayar, "Vision and the atmosphere," *Int. J. Comput. Vis.*, vol. 48, no. 3, pp. 233–254, 2002.
4. Joung-Youn Kim, Lee-Sup Kim, and Seung-Ho Hwang, "An advanced contrast enhancement using partially overlapped sub-block histogram equalization," *IEEE Trans. Circuits Syst. Video Tech.*, vol. 11, no. 4, pp. 475–484, 2001.
5. S. G. Narasimhan and S. K. Nayar, "Chromatic framework for vision in bad weather," *Proc. IEEE Conf. Comput. Vis. Pattern Recognit. (CVPR)*, vol. 1, pp. 598–605, Jun. 2000.
6. S. G. Narasimhan and S. K. Nayar, "Contrast restoration of weather degraded images," *IEEE Trans. Pattern Anal. Mach. Learn.*, vol. 25, no. 6, pp. 713–724, Jun. 2003.
7. J. Kopf et al., "Deep photo: Model-based photograph enhancement and viewing," *ACM Trans. Graph.*, vol. 27, no. 5, pp. 116:1–116:10, Dec. 2008.
8. S. G. Narasimhan and S. K. Nayar, "Interactive deweathering of an image using physical models," *Proc. IEEE Workshop Color Photometric Methods Comput. Vis., Nice, France*, vol. 6, no. 4, pp. 1–7, Oct. 2003.
9. Z. Xu, X. Liu, and X. Chen, "Fog removal from video sequences using contrast limited adaptive histogram equalization," *Proc. Int. Conf. Comput. Intell. Softw. Eng. (CISE)*, pp. 1–4, Dec. 2009.
10. Z. Mi, H. Zhou, Y. Zheng, and M. Wang, "Single image dehazing via multi-scale gradient domain contrast enhancement," *IET Image Process.*, vol. 10, no. 3, pp. 206–214, 2016.
11. C. O. Ancuti and C. Ancuti, "Single image dehazing by multi-scale fusion," *IEEE Trans. Image Process.*, vol. 22, no. 8, pp. 3271–3282, Aug. 2013.
12. Y. Li, Q. Miao, R. Liu, J. Song, Y. Quan, and Y. Huang, "A multi-scale fusion scheme based on haze-relevant features for single image dehazing," *Neurocomputing*, vol. 283, pp. 73–86, Mar. 2018.
13. J. Zhou and F. Zhou, "Single image dehazing motivated by Retinex theory," *Proc. 2nd Int. Symp. Instrum. Meas., Sensor Netw. Automat. (IMSNA)*, pp. 243–247, Dec. 2013.

14. J. Wang, K. Lu, J. Xue, N. He, and L. Shao, "Single image dehazing based on the physical model and MSRCR algorithm," *IEEE Trans. Circuits Syst. Video Technol.*, vol. 28, no. 9, pp. 2190–2199, Sep. 2018.

15. R. T. Tan, "Visibility in bad weather from a single image," *Proc. IEEE Conf. Comput. Vis. Pattern Recognit. (CVPR)*, Anchorage, AK, pp. 1–8, Jun. 2008.

16. R. Fattal, "Single image dehazing," *ACM Trans. Graph.*, vol. 27, no. 3, p. 72, Aug. 2008.

17. K. He, J. Sun, and X. Tang, "Single image haze removal using dark channel prior," *IEEE Trans. Pattern Anal. Mach. Intell.*, vol. 33, no. 12, pp. 2341–2353, 2011.

18. Q. Zhu, J. Mai, and L. Shao, "A fast single image haze removal algorithm using color attenuation prior," *IEEE Trans. Image Process.*, vol. 24, no. 11, pp. 3522–3533, 2015.

19. D. Berman, T. Treibitz, and S. Avidan, "Single image dehazing using Haze-Lines," *IEEE Trans. Pattern Anal. Mach. Intell.*, vol. 42, no. 3, pp. 720–734, 2018.

20. R. He, Z. Wang, Y. Fan, and D. D. Feng, "Multiple scattering model based single image dehazing," *Proc. IEEE 8th Conf. Ind. Electron. Appl. (ICIEA)*, pp. 733–737, Jun. 2013.

21. M. Ju, Z. Gu, and D. Zhang, "Single image haze removal based on the improved atmospheric scattering model," *Neurocomputing*, vol. 260, pp. 180–191, Oct. 2017.

22. M.-Y. Ju, C. Ding, D.-Y. Zhang, and Y. J. Guo, "Gamma-correction-based visibility restoration for single hazy images," *IEEE Signal Process. Lett.*, vol. 25, no. 7, pp. 1084–1088, Jul. 2018.

23. K. Tang, J. Yang, and J. Wang, "Investigating haze-relevant features in a learning framework for image dehazing," *Proc. IEEE Conf. Comput. Vis. Pattern Recognit. (CVPR)*, Jun. 2014, pp. 2995–3002.

24. M. Kim, S. Yu, S. Park, S. Lee, and J. Paik, "Image dehazing and enhancement using principal component analysis and modified haze features," *Appl. Sci.*, vol. 8, no. 8, p. 1321, 2018.

25. B. Cai, X. Xu, K. Jia, C. Qing, and D. Tao, "DehazeNet: An end-to-end system for single image haze removal," *IEEE Trans. Image Process.*, vol. 25, no. 11, pp. 5187–5198, Nov. 2016.

26. W. Ren, S. Liu, H. Zhang, J. Pan, X. Cao, and M.-H. Yang, "Single image dehazing via multi-scale convolutional neural networks," *Proc. ECCV*, pp. 154–169, 2016.

27. B. Li, X. Peng, Z. Wang, J. Xu, and D. Feng, "AOD-Net: All-in-one dehazing network," *Proc. IEEE Int. Conf. Comput. Vis. (ICCV)*, pp. 4780–4788, Oct. 2017.

28. G. Buchsbaum, "A spatial processor model for object colour perception," *J. Franklin Inst.*, vol. 310, no. 1, pp. 1–26, 1980.

29. G. Finlayson and E. Trezzi, "Shades of gray and colour constancy," *Proc. 12th Color Imag. Conf.*, pp. 37–41, 2004.

30. J. van de Weijer, T. Gevers, and A. Gijsenij, "Edge-based color constancy," *IEEE Trans. Image Process.*, vol. 16, no. 9, pp. 2207–2214, Sep. 2007.

31. E. J. Mccartney and F. F. H. Jr, "Optics of the atmosphere: Scattering by molecules and particles," *Optica Acta Int. J. Optics*, vol. 14, no. 9, pp. 698–699, 1977.

3 FishNet: Automated Fish Species Recognition Network for Underwater Images

Monika Mathur and Nidhi Goel
IGDTUW

CONTENTS

3.1 INTRODUCTION

Classification of fish species has recently obtained the focus in underwater image processing as monitoring the health, population and behavior of different fish species [1]. It is very important for maintaining the status of their natural habitats and marine ecological system [2–4]. Fish species classification is among the toughest tasks for researchers and marine scientists due to underwater environment involving absorption and scattering of light [5]. It leads to dull and hazy images of underwater scenario causing hindrances to classification of fish species [6, 7].

Visual classification of fish species is a complex and time-consuming task that involves destructive measures like capturing and killing of fishes for visual census by deep sea divers. Manual classification is also dangerous for divers as some species are poisonous and can be life threatening for them. To avoid these destructions for both humans and fishes, an automatic system is the need of the hour which can classify the fish species precisely without human interference [8]. Deep learning or machine

DOI: 10.1201/9781003138068-3

learning is the emerging area which is being used in many applications for classification, recognition, detection, etc., [9] and can also be used for fish species classification in present scenario.

This chapter proposes two automatic underwater fish classification networks with high accuracy. Deep neural networks can be used as automatic system for classification problem but training a deep neural network from initial point requires large data set of underwater images containing different fish species as it has large weights for training. But the data sets available for fish classification are of small size which are not sufficient for training the deep neural networks. So, transfer learning can be the best solution in such scenarios where scarce data sets are obstacles for training of deep neural networks. Transfer learning uses only last few layers of pre-trained network for classification and is thus fast and also requires less data for training.

The proposed chapter is organized in the following sequence: Section 3.2 surveys the related work and Section 3.3 explains the proposed networks along with architecture of AlexNet and ResNet50. Section 3.4 analyzes the result and shows comparative analysis. Section 3.5 concludes the chapter.

3.2 LITERATURE SURVEY

Various works have been done in the area of fish classification in the last decade [12, 13]. An automatic fish classification method is proposed by Salvo et al. [14] by utilizing the combination of image analysis and artificial immune system techniques. Shape, appearance and motion of the fishes are identified by using SIFT (Scale-Invariant Feature Transform) and PCA (Principal Component Analysis). Clustering of fishes is done by adaptive radius immune technique followed by nearest neighbor algorithm. Cabreira et al. [15] proposed a technique for the automatic classification of fish species depending on their echo recordings. Three parameters, namely, bathymetric, energetic and morphometric were mined from the echo recordings to serve as inputs to the Artificial Neural Networks used for testing. Classification accuracy of 96% was obtained, depending on the type of school parameters and networks utilized. Sayed et al. [16] proposed a fish classification network based on a modified version of crow search optimization algorithm. The network pre-processes the images by applying median filtering for image smoothing and denoising. k-mean clustering is further added for segmentation.

Support vector machine (SVM) and decision trees are used for species classification. The system achieves a classification accuracy of 74% for classification based on fish order and 96% for classification based on classes including Actinopterygii and Chondrichthyes. Boom et al. [17] proposed a heuristics tree-based classification method and finally compares it to a state-of-the-art tree on a live fish image data set. It outperforms the baseline method by achieving an accuracy of 90.0%. Andayani et al. [18] proposed a classification method which includes scaling, cropping, Region of Interest (ROI), gray scale, Hue, Saturation and Value (HSV) color models for image enhancement. Gray level co-occurrence matrix texture features, geographical invariant moment features and HSV color feature extraction methods are used for feature extraction of fish species. Finally classification of these species is done using Probabilistic Neural Network (PNN). Fish species classification method proposed by Castillo et al. [19] utilizes SVM and two supervised Artificial Neural Networks, viz., PNN and Multilayer Perceptron (MLP). Acoustic records have been used as descriptor for the neural networks. The obtained results show that both MLP and SVM methods with accuracy 89.5% outperform the PNN with classification accuracy of 79.4%.

The study proposed by Alsmadi et al. [20] establishes a system based on feature selection and combination theory between major extracted features. This system could classify fishes by utilizing anchor points, texture and statistical measurements. Finally, meta-heuristic algorithm (i.e., Memetic Algorithm) with back-propagation, i.e., MA-B Classifier, is used for generic fish classification of dangerous and non-dangerous species. Dangerous species are further classified as Poison and predatory fishes, whereas non-dangerous fish species are categorized as food and garden family. Accuracy recognition rates of 90% and 82.25% are achieved with proposed MA-B Classifier and the back-propagation algorithm, respectively. Jing et al. [21] proposed an automatic network by combining the fish color and texture with multiclass SVM for fish species identification. Alsmadi et al. [22] also proposed a hybrid fish species classification system with a back propagation and Tabu search, named as Genetic Algorithm with Tabu search with a Back-propagation algorithm) GTB Classifier. It calculates the fish feature by combining the color signature, extracted shape and color texture features. The fishes were classified as poison, predatory, food and garden species. The proposed system was tested for the back-propagation algorithm and GTB classifier with classification rate of 82.1% and 87%, respectively.

Chuang et al. [23] uses unsupervised learning for fish feature extraction and classification of species is done using an error resilient classifier. Sharmin et al. [24] proposed an automatic freshwater fish classification network. The network pre-processes the underwater images by resizing the fish images, histogram equalization and finally conversion to grayscale images. Feature extraction includes geometric measurements, RGB and HSV colors features and GLCM features. SVM is the classifier used for species recognition and achieves a highest accuracy value of 94.2%. Further, Qiu et al. [25] recognizes the fishes in small-scale and low-quality image data set by proposing transfer learning based squeeze and excitation networks. Data augmentation and super-resolution reconstruction are used for enhancing this data set and hence improve upon the accuracy.

Hridayani et al. [26] proposed a VGG16 based network for classification of 50 fish species. The data set is divided into four different types, i.e., RGB image, Canny filter image, simple blending image and blending + RGB image. The results show that data set consisting of blending + RGB images is best by achieving the Genuine Acceptance Rate of 96.4%. Qin et al. [27] proposed Deep-CNN approach on the Fish for knowledge data set consisting of 27,370 fish images and achieved an accuracy of 98.57% in fish classification. Feature extraction of fish images has been done using neural network and PCA in the method proposed by Boom et al. [28]. Block-wise histogram and binary hashing are used for feature pooling and non-linear layers, respectively. Further, SVM classifier performs the final classification.

Most of the above-mentioned papers either use the large data sets to train their network from scratch or uses data augmentation techniques to increase the number of training images to enhance the accuracies artificially. But the proposed methods use the transfer learning-based approach to achieve better accuracies even with limited or small data set. Next section describes the two proposed methods along with describing the transfer learning.

3.3 PROPOSED APPROACH

The present paper proposes transfer learning for fish species classification. Transfer learning is a machine learning technique which builds accurate deep learning models in a timesaving way [29]. In transfer learning, learning process is not done from scratch; rather it extracts the features from pre-trained layers of the network while solving a different problem [30]. In this way transfer learning based networks make use of earlier learnings of network and resist starting it from scratch. Transfer learning generally makes use of pre-trained networks. Pre-trained networks are the networks that are trained on a large data set consisting of millions of images to solve a problem identical to the one that we are interested in. Instead of training a network (deep networks) from scratch which is computationally complex, time-consuming and very costly, it is better to use pre-trained networks (e.g., AlexNet [11], VGG [10], MobileNet [31], GoogleNet [32], ResNet [33] etc.) to save both time and resources.

Canziani et al. [34] presents a comprehensive review of pre-trained networks' performance using images from the ImageNet challenge [35]. The major cause of overfitting and loss of generality in the neural networks is due to lack of high number of data points for the model to learn from. The use of transfer learning compensates for this as the existing state-of-the-art neural networks are already trained on millions of images. These models when fine-tuned for a particular application, for example, classification of fish species in this case, decrease the variance and bias for the model.

There are various available pre-trained neural networks like AlexNet [11], VGG-16 [10], VGG-19 [10], etc., that have already been trained for millions of images. These neural networks have learnt parameters that are achieved by training with millions of images and can also be used for other classification problems, like fish species classification in this paper. Selection of pre-trained network which is both fast and accurate is a tough task. For proper choice of network, comparison among three networks (AlexNet, VGG-16, VGG-19) is done in terms of accuracy, execution time and mini batch loss. For fish classification Queensland University of Technology QUT data set [36] with 10 epochs and 16 batch-size is trained using NVIDIA Quadro K2200 GPU. Table 3.1 shows that AlexNet is both accurate and faster than other two and is a suitable choice for fish species classification.

TABLE 3.1
Comparison of Pre-Trained Networks

	Accuracy (%)	Time (s)	Mini Batch Loss
VGG-16 [10]	73.33	3,038.40	0.1001
VGG-19 [10]	79.17	7,237.76	0.0638
AlexNet [11]	89.17	101.94	0.0327

3.3.1 ARCHITECTURE OF ALEXNET

The neural network defined by Alex et al. [11], also commonly called as AlexNet, is a convolutional neural network with 60 million parameters and 650,000 neurons. It is trained on huge data set of 1.2 million images for classifying 1,000 different categories of objects, animals, etc. AlexNet network has five 2-D convolutional layers. The convolutional layers in the beginning of the network learn low-level features (edges, blobs, colors, etc.) due to small receptive field size, whereas the last layers of the network have larger receptive field size and hence they learn high-level features or task-specific features. The architecture of pre-trained AlexNet has five convolutional layers and three fully connected layers. The first and second convolutional layer has 96 kernels of size $11 \times 11 \times 3$ and 256 kernels of size $5 \times 54 \times 48$, respectively, followed by overlapping softmax layer. Third, fourth and fifth convolutional layers of AlexNet have 384 kernels of size $3 \times 3 \times 256$, 384 kernels of size $3 \times 3 \times 192$ and 256 kernels of size $3 \times 3 \times 192$, respectively. The 2 out of 3 fully connected layers have 4,096 neurons each and the last or classification layer has 1,000 neurons. Convolutional layers are succeeded by max pooling layers. 'Dropout' regularization method is used for avoiding overfitting of fully connected layers because it randomly switches off some percentage of the neurons during training. Figure 3.1 shows the architecture of AlexNet.

3.3.2 FINE TUNING THE ALEXNET NEURAL NETWORK

The last few layers of the neural network need to be modified to apply transfer learning to the neural network. The pre-trained neural networks have learned rich feature representations for a wide variety of images. In the architecture of AlexNet, the last three layers are fully connected layers with last being the classification layer or 1,000 way softmax layer. The softmax layer gives the probability of a test image belonging to a particular class. AlexNet has originally been designed for 1,000 categories. To classify different number of categories, only the last three layers of AlexNet has been changed to fully connected layers with 23 way softmax (or 37-way softmax for QUT data set) instead of 1,000-way softmax layer.

Twenty three is the number of categories in which fish images are to be classified. The network is trained again for learning the weights for the last few modified layers. For transfer learning the network is slightly changed. Learning rates control the amount by which network is changed during training. Proposed architecture does

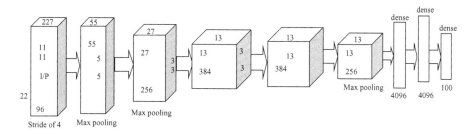

FIGURE 3.1 Architecture of AlexNet.

not modify learning rates (0.001) of the first 22 original layers. These learning rates are already very low and they are not lowered further. The rest of the layers use the weights from the initial training on millions of images. However, before training with the new data set images are resized to 227×227 pixel size as this is what the AlexNet expects for its input. The new fine-tuned network is named as 'FishNet' and Figure 3.2 shows its architecture.

3.3.3   ARCHITECTURE OF RESNET50

The last few layers of the neural network are modified to apply transfer learning to it. The pre-trained neural networks are already trained to learn rich feature representations for various types of images. In the present paper pre-trained network, ResNet50 is used for transfer learning [33]. ResNet50 stands for Residual Network and is used for many computer vision tasks. This model was the winner of ImageNet challenge-2015. The important characteristic of ResNet50 is to train extremely deep neural networks even with more than 150 layers successfully by overcoming the problem of vanishing gradients.

The ResNet50 architecture consists of four stages as shown in Figure 3.3. This network accepts the images with width and height as multiples of 32. The present paper uses the input size as 224×224×3, with 3 as number of color channels. ResNet50 architecture initially starts with convolution and max-pooling using 7×7 and 3×3 kernel sizes, respectively. Stage 1 of ResNet

network referred in Figure 3.4 has three residual blocks and each block comprises three layers with kernel size 64, 64 and 128 to perform the convolution operation in stage 1.

The curved arrows in Figure 3.3 show the identity connection. The dashed arrows in Figure 3.4 show the convolution operation in the Residual Block with stride 2. This reduces input size, i.e., height and width to half and increases channel width to double. Input size keeps on decreasing to half and channel width keeps on increasing to twice as process progresses from one stage to next, i.e., from stage 1 to stage 4. Finally, ResNet50 has an Average Pooling layer followed by a fully connected layer having 1,000 neurons.

3.3.4   RESNET50 AS FEATURE EXTRACTOR (FISH _ NET _ SVM)

Fish_Net_SVM also extract image features from ResNet50 network and use those learned features to train an image classifier known as SVM. Pre-trained ResNet50 network is trained on a very large data set consisting of millions of images and can classify

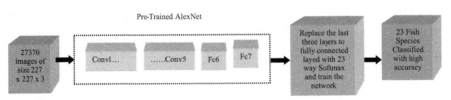

FIGURE 3.2   Fine tuning of AlexNet.

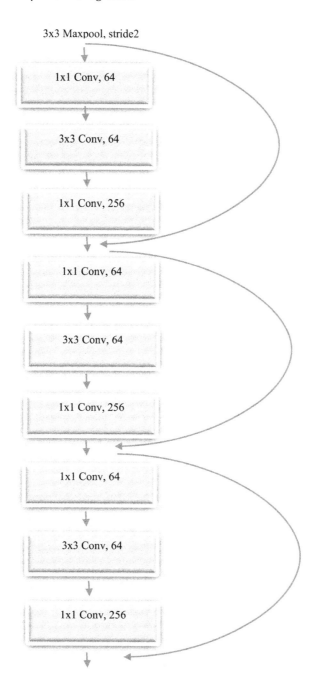

To Stage 2

FIGURE 3.3 Architecture of ResNet50.

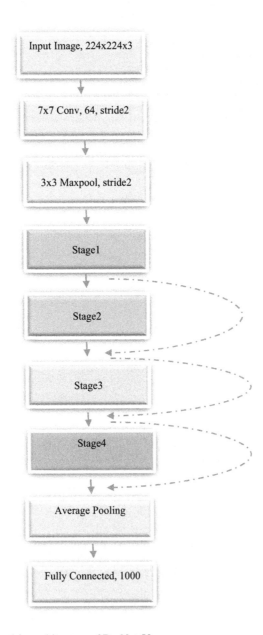

FIGURE 3.4 Stage1 in architecture of ResNet-50.

images into thousand categories, like cap, pencil, torch, animals, birds, insects, etc. Due to large data set ResNet50 has learned feature representations for different types of images. These feature representations of input images are arranged hierarchically.

Initial layers contain low-level features like dots, lines, curves, etc., but deeper layers contain higher level features like edges, contours, etc. Feature representations of the training and test images are achieved using activations on the global pooling layer

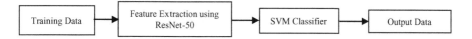

| Training Data | → | Feature Extraction using ResNet-50 | → | SVM Classifier | → | Output Data |

FIGURE 3.5 ResNet50 as feature extractor.

(avg_pool) at the end of the ResNet50 network. The avg_pool layer pools these features over all spatial locations and gives a total of 2,048 features. The class labels are extracted from both the training and test data sets. Features extracted from the training images act as predictor variables for a multi-class SVM for classification. Block diagram of Fish_Net_SVM is shown in Figure 3.5.

3.4 RESULTS

3.4.1 DATA SET

Effectiveness of the proposed CNN network FishNet is measured on fish data set built by the experiments and is named as Fish for Knowledge (F4K) [28], where 23 fish species are physically categorized by marine biologists. Of 27,370 fish images 80%, i.e., 21,896 images, are used for training, whereas 20%, i.e., 5,474 images, are used for testing data.

The QUT fish data set comprises 3,960 images collected from 468 species in under controlled, out-of-the-water and underwater conditions. For proposed methods 37 species have been chosen for classification comprising 600 images in natural underwater condition. 70:30 ratio has been used for training and test images. The proposed methods have been tested on these large (F4K) as well as small (QUT) data sets.

Matlab 2018a is used for implementation of algorithms and evaluation of results. Figure 3.6 shows examples of the 23 species to be classified whereas Table 3.2 shows the distribution of fishes in each species.

3.4.2 RESULT EVALUATION AND COMPARISONS

Both the proposed methods (FishNet and Fish_Net_SVM) are evaluated using F4K and QUT data sets. F4K has 27,370 images and is called as large data set whereas QUT has 600 images and is called as small data set. FishNet and Fish_Net_SVM with large data set give an accuracy of 98.67% and 95.90%, respectively. Time consumed by FishNet is 1,930 min and 5 s whereas the time taken by Fish_Net_SVM is 662.8795 s for 5 epochs and 0.001 learning rate. The above mentioned results depict that FishNet (fine-tuning the pre-trained network according to our data) is better if accuracy is priority and Fish_Net_SVM (Feature extraction from pre-trained network + SVM classifier) is better if time consumption or speed of the network is the main concern.

Similarly, the results are also evaluated for small data set to measure the changes in the accuracy and speed. As the data are small both the accuracy and time consumed are reduced for both the methods. FishNet and Fish_Net_SVM with small data set give an accuracy of 89.17% and 77.86%, respectively. Time consumed by FishNet is 21 min and 35 s whereas the time taken by Fish_Net_SVM is 30.32 s for 5 epochs and 0.001 learning rate. The results of both the methods are tabulated in Table 3.3.

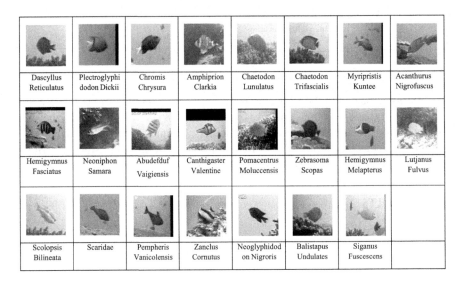

Dascyllus Reticulatus	Plectroglyphidodon Dickii	Chromis Chrysura	Amphiprion Clarkia	Chaetodon Lunulatus	Chaetodon Trifascialis	Myripristis Kuntee	Acanthurus Nigrofuscus
Hemigymnus Fasciatus	Neoniphon Samara	Abudefduf Vaigiensis	Canthigaster Valentine	Pomacentrus Moluccensis	Zebrasoma Scopas	Hemigymnus Melapterus	Lutjanus Fulvus
Scolopsis Bilineata	Scaridae	Pempheris Vanicolensis	Zanclus Cornutus	Neoglyphidodon Nigroris	Balistapus Undulates	Siganus Fuscescens	

FIGURE 3.6   Examples of fish species from F4K data set.

TABLE 3.2
F4K Data Set with Fish Species Distribution

Fish Species	No. of Samples	Fish Species	No. of Samples
Amphiprion clarkia	4,049	*Dascyllus recticulatus*	12,112
Chaetodon lunulatus	2,534	*Chromis chrysura*	3,593
Myripristis kuntee	450	*Plectroglyphidodo dickii*	2,683
Neoniphon samara	299	*Lutjanus fulvus*	206
Acanthurus nigrofuscus	218	*Hemigymnus fasciatus*	241
Chaetodon trifascialis	190	*Canthigaster valentine*	147
Pomacentrus moluccensis	181	*Scolopsis bilineata*	49
Abudefduf vaigiensis	98	*Zebrasoma scopas*	90
Zanclus cornutus	21	*Scaridae*	56
Balistapus undulates	41	*Hemigymnus melapterus*	42
Siganus fuscescens	25	*Pempheris vanicolensis*	29
Neoglyphidodon nigroris	16		

TABLE 3.3
Comparative Analysis of FishNet and Fish_Net_SVM in Terms of Accuracy and Time Consumption

	FishNet		Fish_Net_SVM	
	Accuracy (%)	Time	Accuracy (%)	Time (s)
QUT data set	89.17	21 min 35 s	77.86	30.32
F4K data set	98.67	1,930 min	95.90	662.8795

The results of 23 fish species classification using FishNet method are shown in Table 3.4 in terms of accuracy along with its comparison to existing methods on the Fish4Knowledge data set.

Various machine learning tools and techniques have been used as reference methods for comparisons. Methods based on LDA as feature extractor and SVM as classifier [37] obtained an accuracy of 80.14% by using fish images without background. A validation accuracy of 89.79% has been achieved by nearest neighbor method for fish species classification. A simple SVM classifier-based approach obtains an accuracy of 82.98% by training the model on the raw pixels.

Further, softmax classifier [39] in DeepFish model achieves an accuracy of 87.56% whereas VLFeat method [38] used for comparison gives an accuracy of 93.58%. Results from DeepFish architecture [39] with and without data augmentation give a test accuracy of 98.23% and 98.59%. The comparison results are listed in Table 3.4. Deep-CNN [27] uses deep learning architecture for designing its three convolutional layers. It is trained from scratch for fish and plankton classification. It achieves an accuracy of 98.57%. Alex-FT-Soft and Alex-SVM [40] both use AlexNet for feature extraction, whereas softmax and SVM classifiers for fish recognition and give an accuracy of 96.61% and 98.57%, respectively.

FishNet also extracts features from fish images using the pre-trained AlexNet network and classifies them by fine-tuning. The validation or test accuracy of 98.67% demonstrates that the proposed network with transfer learning outperforms other state-of-the-art methods for fish species classification. 20 epochs in 9,630 s for FishNet further indicate that transfer learning-based neural networks are much faster as there is no need to train them for as many epochs as a new

TABLE 3.4
Comparative Analysis of FishNet and Fish_Net_SVM with Various Models on F4K Data

Classification Models	Accuracy (%)
Raw-Pixel Nearest Neighbor [37]	89.79
Raw-Pixel SVM [37]	82.92
LDA + SVM [37]	80.14
VLFeat Dense-SIFT [38]	93.58
Raw-Pixel Softmax [39]	87.56
DeepFish-SVM [39]	98.23
DeepFish-SVM-Aug-scale [39]	98.64
DeepFish-Softmax-Aug [39]	92.55
DeepFish-SVM-Aug [39]	98.59
Alex-FT-Soft [40]	96.61
DeepFish-Softmax-Aug-scale [39]	98.49
Deep-CNN [27]	98.57
Alex-SVM [40]	98.57
Fish_Net_SVM	95.90
FishNet	98.67

model would require [41], i.e., to train from beginning. Transfer learning-based models also require comparatively less images (only 27,370 images for FishNet) as compared to huge data sets (million or more) required for training a model from ground level [32]. Hence, FishNet model is accurate and fast solution for fish categorization with minimum labor.

3.5 CONCLUSION

Automatic fish classification is an important task for marine biologists for studying the behavior and traits of various fish species. Traditional methods are destructive and possess threat to the marine life by physically capturing them. These methods also demand time and labor costs but transfer learning-based FishNet and Fish_Net_SVM provide a cost-effective and efficient solution to the above-mentioned problem. The proposed FishNet and Fish_Net_SVM networks for fish classification do not require a very large data set to get higher test accuracy. In the scenarios, where availability of sufficient data is an issue, transfer learning proves to be a better solution as pre-trained networks are used with fine tuning according to desired requirements. The proposed FishNet network extracts fish features from images by using learned weights of the pre-trained AlexNet network and it uses softmax layer to classify fish species. On the other hand Fish_Net_SVM network extracts image features from ResNet50 network and uses those learned features to train an image classifier known as SVM. Result analysis highlights the effectiveness of FishNet and Fish_Net_SVM by outperforming the state-of-the-art fish classification models. Hence, the obtained results also indicate that computation complexity and accuracy are better for the transfer-learned models as compared to models trained from scratch.

REFERENCES

1. R. Schettini and S. Corchs. Underwater image processing: State of the art of restoration and image enhancement methods. *EURASIP Journal on Advances in Signal Processing*, 2010(1):746052, 2010.
2. S. Hasija, M. J. Buragohain, and S. Indu. Fish species classification using graph embedding discriminant analysis. In *2017 International Conference on Machine Vision and Information Technology (CMVIT)*, pages 81–86, 2017.
3. S. Marini, E. Fanelli, and V. Sbragaglia. Tracking fish abundance by underwater image recognition. *Scientific Reports*, 8:1–12, 2018.
4. N. Andrialovanirina, D. Ponton, F. Behivoke, J. Mahafina, and M. Leopold. A powerful method for measuring fish size of small-scale fishery catches using ImageJ. *Fisheries Research*, 223:105425, 2020.
5. M. Mathur and N. Goel. Enhancement of non-uniformly illuminated underwater images. *International Journal of Pattern Recognition and Artificial Intelligence* 35:3, 1–22, 2020.
6. M. Mathur and N. Goel. Enhancement of underwater images using white balancing and Rayleigh-stretching. In *5th International Conference on Signal Processing and Integrated Networks (SPIN)*, pages 924–929, 2018.
7. M. Mathur and N. Goel. Dual domain approach for colour enhancement of underwater images. In *11th Indian Conference on Computer Vision, Graphics and Image Processing*, pages 1–6, 2018.

8. M. Mathur, D. Vasudev, S. Sahoo, and N. Goel. Crosspooled fishnet: Transfer learning based fish species classification model. *Multimedia Tools and Application*, 79:31625–31643, 2020.

9. S. Kaur and N. Goel. A dilated convolutional approach for inflammatory lesion detection using multi-scale input feature fusion. In *IEEE International Conference on Multimedia Data*, 2020.

10. K. Simonyan and A. Zisserman. Very deep convolutional networks for large-scale image recognition. *CoRR*, abs/1409.1556:1–14, 2014.

11. A. Krizhevsky, I. Sutskever, and G. E. Hinton. ImageNet classification with deep convolutional neural networks. *Communication ACM*, 60(6):84–90, 2017.

12. M. K. Alsmadi and I. Almarashdeh. A survey on fish classification techniques. *Journal of King Saud University - Computer and Information Sciences*: 1–14, 2020.

13. R. F. Syreen and K. Merriliance. A survey on underwater fish species detection and classification. *International Journal of Computer Sciences and Engineering*, 7:95–98, 2019.

14. C. Spampinato, D. Giordano, R. D. Salvo, Y. H. Jessica, C. Burger, R. B. Fisher, and G. Nadarajan. Automatic fish classification for underwater species behavior understanding. In *ARTEMIS@ACM Multimedia*, 2010.

15. A. Cabreira, M. Tripode, and A. Madirolas. Artificial neural networks for fish species identification. *ICES Journal of Marine Science*, 66:6, 2009.

16. G. I. Sayed, A. E. Hassanien, A. Gamal, and H. A. Ella. An automated fish species identification system based on crow search algorithm. The International Conference on Advanced Machine Learning Technologies and Applications (AMLTA2018). *Advances in Intelligent Systems and Computing*, 723:112–123, 2018.

17. P. X. Huang, B. J. Boom, and B. Fisher. Hierarchical classification with reject option for live fish recognition. *Machine Vision and Applications*, 26:89–102, 2015.

18. U. Andayani, A. Wijaya, R. Rahmat, B. Siregar, and M. Syahputra. Fish species classification using probabilistic neural network. *Journal of Physics: Conference Series*, 1235:012094, 2019.

19. H. Robotham, P. Bosch, J. C. Guti_errez-Estrada, J. Castillo, and I. Pulido-Calvo. Acoustic identification of small pelagic fish species in Chile using support vector machines and neural networks. *Fisheries Research*, 102(1):115–122, 2010.

20. M. Alsmadi, M. Tayfour, R. Alkhasawneh, U. Badawi, I. Almarashdeh, and F. Haddad. Robust features extraction for general fish classification. *International Journal of Electrical and Computer Engineering (IJECE)*, 9:51–92, 2019.

21. H. Jing, L. Daoliang, D. Qingling, H. Yueqi, C. Guifen, and S. Xiuli. Fish species classification by color, texture and multi-class support vector machine using computer vision. *Computers and Electronics in Agriculture*, 88:133–140, 2012.

22. M. Alsmadi. Hybrid genetic algorithm with Tabu search with back-propagation algorithm for fish classification: Determining the appropriate feature set. *International Journal of Applied Engineering Research*, 14(23):4387–4396, 2019.

23. M. Chuang, J. Hwang, and K. Williams. A feature learning and object recognition framework for underwater fish images. *IEEE Transactions on Image Processing*, 25(4):1862–1872, 2016.

24. I. Sharmin, N. Islam, I. Jahan, T. Joye, R. Rahman, and M. Habib. Machine vision based local fish recognition. *SN Applied Sciences*, 1:1–12, 2019.

25. C. Qiu, S. Zhang, C. Wang, Z. Yu, H. Zheng, and B. Zheng. Improving transfer learning and squeeze- and-excitation networks for small-scale fine-grained fish image classification. *IEEE Access*, 6:78503–78512, 2018.

26. P. Hridayami, I. K. G. D. Putra, and K. S. Wibawa. Fish species recognition using vgg16 deep convolutional neural network. *Journal of Computing Science and Engineering*, 13(3):124–130, 2019.

27. H. Qin, X. Li, Z. Yang, and M. Shang. When underwater imagery analysis meets deep learning: A solution at the age of big visual data. In *OCEANS 2015-MTS/IEEE Washington*, pages 1–5, 2015.

28. B. J. Boom, P. X. Huang, C. Spampinato, and S. Palazzo. Fish recognition ground-truth data. http://groups.inf.ed.ac.uk/vision/Fish4Knowledge/Website/Groundtruth/Recog/, 2013. [Online; accessed August 2020].

29. W. Rawat and Z. Wang. Deep convolutional neural networks for image classification: A comprehensive review. *Neural Computation*, 29:1–98, 2017.

30. C. Tan, F. Sun, T. Kong, W. Zhang, C. Yang, and C. Liu. A survey on deep transfer learning. In *Artificial Neural Networks and Machine Learning -ICANN 2018*, pages 270–279, 2018.

31. A. G. Howard, M. Zhu, B. Chen, D. Kalenichenko, W. Wang, T. Weyand, M. Andreetto, and H. Adam. MobileNets: Efficient convolutional neural networks for mobile vision applications, 1–9, arXiv:1704.04861 **[cs.CV]**, 2017.

32. C. Szegedy, W. Liu, Y. Jia, P. Sermanet, S. Reed, D. Anguelov, D. Erhan, V. Vanhoucke, and A. Rabinovich. Going deeper with convolutions. In *2015 IEEE Conference on Computer Vision and Pattern Recognition (CVPR)*, pages 1–9, 2015.

33. K. He, X. Zhang, S. Ren, and J. Sun. Deep residual learning for image recognition. In *2016 IEEE Conference on Computer Vision and Pattern Recognition (CVPR)*, pages 770–778, 2016.

34. A. Canziani, A. Paszke, and E. Culurciello. An analysis of deep neural network models for practical applications. arXiv:1605.07678 [cs.CV], 2016.

35. J. Deng, W. Dong, R. Socher, L. Li, K. Li, and L. Fei-Fei. Imagenet: A large-scale hierarchical image database. In *2009 IEEE Conference on Computer Vision and Pattern Recognition*, pages 248–255, 2009.

36. Z. Y. Ge, C. McCool, and P. Corke. Fish dataset. https://wiki. qut.edu.au/display/cyphy/Fish+Dataset, 2016. [Online; accessed August 2020].

37. C. C. Chang and C. Jen. Libsvm: A library for support vector machines. *ACM Transaction on Intelligent System Technology*, 2(3):1–27, 2011.

38. A. Vedaldi and B. Fulkerson. Vlfeat: An open and portable library of computer vision algorithms. In *Proceedings of the 18th ACM International Conference on Multimedia*, pages 1469–1472, 2010.

39. H. Qin, X. Li, J. Liang, and C. Zhang. DeepFish: Accurate underwater live fish recognition with a deep architecture. *Neurocomputing*, 187:49–58, 2016.

40. A. B. Tamou, A. Benzinou, and K. Nasreddine. Underwater live fish recognition by deep learning. In *Image and Signal Processing*, pages 275–283, 2018.

41. M. D. Zeiler and R. Fergus. Visualizing and understanding convolutional networks. In *Computer Vision - ECCV 2014*, pages 818–833, 2014

4 Person Identification in UAV Shot Videos by Using Machine Learning

Manju Khari, Renu Dalal, Arti Sharma,
and Bhavya Mehta
Ambedkar Institute of Advanced Communication
Technologies & Research

CONTENTS

4.1   INTRODUCTION

Unmanned aerial vehicles (UAVs) or so-called drones can easily reach those locations which are too difficult to reach and access and dangerous for human beings or endanger the lives. UAVs can collect images from birds' view through aerial capturing of the surroundings. For observation, remote-controlled military drones (or aircrafts, UAVs) are used in recent years but working with this approach only will not lead to improvements in the research area of drone technology. Research demonstrates that there are several factors involved in the person identification process. If the target is unfamiliar and the limit of the targets is exceeded then the images captured through UAVs have poor quality. Also, movement of targets leads to poor quality of the captured images or videos, pose variations, variation in motions, high and low illumination, altitude, resolution level and background variations make impact on the results efficiency. These can be considered as the gaps between gallery and probes domain. UAVs not only provide accessibility to those areas but it also facilitates monitoring for required time periods. For those areas where human life could be in danger, UAVs with their small size, wide view of field and remote-control abilities make it possible to reach those areas without the physical presence of humans.

In the past decades it has been noticed that researchers are focused on detection, tracking and monitoring in this field. With the advancement and the need of reliable solutions, it needs to focus more on person identification. In this chapter, a data set which has 200 videos with 58 subjects in 411K frames. These videos have been shot from different angles for giving a reliable approach. This data set is based on badge detection ideology. This data set is intended to provide a platform for research in person identification using drones. The videos in this data set have high resolution and pixels. Research demonstrates that there are several factors involved in the person identification process. If the target is unfamiliar and the limit of the targets is exceeded then the images captured through UAVs have poor quality. Also, movement of targets leads to poor quality of the captured images or videos, pose variations, change set in motions, high and low illumination, altitude, resolution level and background variations make an impact on the results efficiency. These can be considered as the gaps between gallery and probes domain.

Considering the areas which are inaccessible for humans in an easy way UAVs or so-called drones can make a great impact by making it easy for humans to reach those particular areas. UAVs not only provide accessibility to those areas but it also facilitates monitoring for required time periods. For those areas where human life could be in danger, UAVs with their small size, wide view of field and remote-control abilities make it possible to reach those areas without the physical presence of humans. In the past decades it has been noticed that researchers are focused on detection, tracking and monitoring in this field. With the advancement and the need of reliable solution, we need to focus more on person identification.

Many applications may come into highlight with the use of identification of persons in UAV shot videos; for example, it can provide application of surveillance, searching of people and also mobile monitoring. Mentioned application can make great impact in future development as face recognition technologies can be employed in UAVs itself.

This technology is very well used in face detection of culprits like finding face of terrorists in unsafe areas and in situations like COVID-19 for tracking crowded places. Many countries have been using drones in trouble zoned areas for identifying the cause of the situation safely. The Central government of India has also used drones at the time of Citizenship Amendment Act (CAA) protest which made it easy for the police to monitor the situation. Also, drones are helpful for disaster management as it can track those areas which are inaccessible by humans. In this chapter, algorithms are compared for face detection and face recognition.

For face detection, two different algorithms are used. One is Viola Jones which needs a full-frontal view of the face and another is Tiny faces which is used to detect faces in crowded places. In this way, we cover each scenario from a single person to a crowded place.

For face recognition, four algorithms are used: (i) Histogram of Oriented Gradients (HOG), (ii) Local Binary Pattern (LBP), (iii) VGG-Face and (iv) Commercial-Off-The-Shelf system (COTS).

Some of the major challenges that we have observed in face recognition is when the drone shot videos are in low resolution. Also, when the distance between the face and the drone is large and varying, it becomes hard for the model to recognize or even detect the face. There are multiple different algorithms for face detection and face recognition but due to the above challenges; these do not give good results on our data set. For improving the current scenario on person identification in UAV shot videos, it has been investigated that how altitude and distance of a person from a drone are going to impact the performance.

While performing different algorithms as discussed above it has been noticed that in results of frame wise identification, VGG-Face algorithm performed best among others with highest accuracy in identification. COTS gave the worst performance with lowest accuracy and LBP model drew the midline of both of these results [1, 2]. The HOG algorithm also performed at a moderate level in comparison to others.

4.2 RELATED WORK

Machine learning is the science that makes computers work without explicit programming. It has close connections to applied mathematics, providing the field with tools, concept and application areas. Machine learning is often related to data mining, how the next subset deals with the analysis of exploratory data, and is called unsupervised learning. Single-shot Multibox Detector is used in tiny face algorithms [3] for face detection. It uses multi-scale feature maps for detection. It adds a convolutional layer at the end of the base network which helps in decreasing the size progressively. It concatenates the output of a multi-scale feature map at the last layer. It also uses convolutional predictors for detection by using 3×3 convolutional kernels.

In this chapter, the basic aim is to evaluate the effects of various parameters on videos that are shot using drones. While capturing videos from UAV devices there are multiple parameters that directly define the quality of data. And consequently, the performance of model is affected. For example, first parameter, variations in pose of person that are being captured and need to be identified through machine learning algorithms give a criterion to consider it as a performance aspect. Second parameter that contributes to the performance measure is the illumination factor. Illumination factor is concerned with the brightness level and lighting effect in the data that is actually captured using UAV devices. Third factor is to consider the altitude levels and distance between UAV devices and the person that is being captured. Last but not the least factor that needs to be focused is the database that is used to train our model and to observe the performance of the model. The data may be familiar or unfamiliar to the model. These are certain challenges that were observed while going through the previous studies that have been implemented in this domain.

Basically, recognition has been applied from a benchmark methodology in previous studies that compares a variety of color—shape image characteristics to find potential compared to a database of a query individual. This method is strengthened by integrating controller [1] suggestions in three different ways. Second, characteristics are measured indicating the degree of conformity (similarity) with input from the controller. Second, on the basis of controller inputs, a classification model is educated. Deep learning models, like convolutionary artificial neural networks (CNNs), may be used to support different facets of the moving and targeting phase to enable one, maybe more drones also to be controlled at once by a person. However, since computing capacity and data for a particular exist, it is not obvious to use machine learning for drones. An arrangements-based technique [2] for learning compact fully connected structures is suggested in this work.

The cross-identification (cross-id) [3] topic of individuals has currently drawn the interest of the computer vision applications due to its critical role in current monitoring systems. In addition, the excellent success in the image recognition task of advanced CNNs made CNN among the most powerful computer vision resources. The output drive was activated and investigators were encouraged to obtain and publish further detailed post-id data. In nearly every current activity recognition study, a basic premise is that images are in mean position, requiring image and cross-camera characteristics to be modelled in place to enable verification. They introduce an experimental analysis throughout this [4] chapter that moves classification in another position: cross-identification on a smartphone, including a robot. They are formalizing certain versions of the traditional facial expression recognition paradigm which are applicable to wireless manipulation.

An important area of study with promising prospects for the film industry [5] is automated Aircraft production design. It aims to massively improve UAV training for diverse uses, while lowering prices dramatically due to conventional training. The overall issue just has not, nevertheless, been clearly established as well as the problems resulting from existing regulations and technological limitations have still not been thoroughly identified. A full description of the automated production design problems is required.

In this section the first part gives an idea of the work that has been done till now in the field of person identification in UAV shot videos and the second part depicts the studies done in deep learning models for person identification. Figure 4.1 presents Variation in Altitude for UAV shot videos. To make the person identification technique more reliable, there is a need to focus on the challenges that are arising in this field. Many studies had been done in image surveillance for object detection. Years later, many other researchers focused on data driven applications. Applications based on the internet of things technology are being proposed for agricultural areas [6, 7]. Deep learning technology is making an impact in the control root systems of flights and giving the facility of next level like vision assistance. One of such systems is PIXHAWK; this is very popular and is based on the capability of computer vision that can perform object recognition and also provides facility of autopilot mode for drone operations.

Person recognition can be understood as providing the information about the person that is being captured in videos. The process to identify a person done in two modules is named as classification followed by localization. There is a major challenge in identification to locate all the people that are present in the frame at a particular instance. It has been noticed that with the advancement of technology human efforts are getting less privileged. Person identification can reduce human

FIGURE 4.1   Variation in altitude.

surveillance to a noticeable level. Detection in the recent years shows the results in the form of confidence scores with some labels in rectangular form. The very first model that was approached for object detection in deep learning was named as Overleaf Network that makes use of CNN. Then after that R-CNN came into light followed by some more approaches like YOLO, R-FCN and SSD [8–10]. All these approaches are made to improve the accuracy in detection models so that we can match the real-world statistics in mobile platforms as well.

Till now various papers had presented approaches for person identification in UAV videos. While these approaches format, here the lead purpose was for interactive operation where operator feedback over multiple iterations was important.

4.2.1 Data Set

Person identification in UAV shot videos is a challenging problem having some factors like movement of UAV device, quality of footage, target person and also some environmental factors. Suppose the scenario where target has unrestricted mobility and can move at any distance from device, then in that case resolution and pose variation will lead to poor quality. Using UAVs as the mainstream source of video is very challenging. Choosing a data set for the particular approach is as important as applying models because data sets play a major role in deciding the efficiency, correctness and reliability of the proposed approach.

Surveillance and monitoring in UAV shot videos have certain restrictions from certain factors like the movement of person or object, quality of shot video, identification of particular person also last but not the least some environmental factors. Different challenges may arise in different scenarios like speed of movement of person, altitude variation and illuminous factor. Keeping all these factors in consideration this data set (http://www.svcl.ucsd.edu/projects/dronefollow/) is used.

To overcome the challenges in person identification using UAV shot videos, an efficient data set is selected. This data set is actually based on frame badge detection, in which the target is wearing a badge that is programmed. These badges integrate two modules in visual pattern. First component is static and stable. These components are used to locate the user, control distance and height between target and UAV device. In this data set data have been collected for different height and illumination levels. Also, there are some images collected at multiple distances.

4.3 PROPOSED WORK

In the very first step, pre-processing of the data has been done so that it increases efficiency by some level. The images and videos which had noise and had low resolution were removed. Then normalized all the face images to identical lighting conditions. Now, reduction in resolution variation in different videos to increase the final accuracy. Face components like eyes, nose, mouth, etc., are extracted for detection and recognition. It is important to extract the features for initial processes like face tracking and recognition. This meaningful data extracted from face image will help to detect a person. The data set is divided into two parts where 30% data as a testing data set and the rest is considered as training data. Flowchart for proposed work is shown in Figure 4.2.

Training data and testing are independent from each other and test data are unseen by the model which increases the difficulty. Two algorithms have been used to detect faces which are viola jones and tiny faces to cover all the lighting conditions. These two approaches cover every situation of data set which helps in increasing the accuracy. In this approach, we computed a baseline for person detection with the help of two face detectors. These detectors are named as Viola Jones and Tiny faces. Both models demonstrate recall and precision values with respect to efficiency [11–13]. For the detection of faces in crowds with lower resolution, here the model named Tiny faces is already

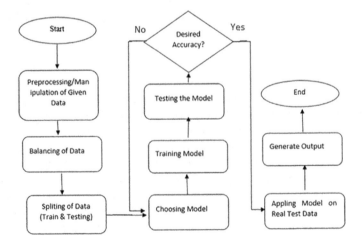

FIGURE 4.2 Flowchart of the proposed work.

pre-trained. Several methods can be used to check the performance of a model like through confusion matrix by calculating. F1 score, mean absolute error, etc., are used for evaluation. In this chapter, we measured the performance by giving scores to each box that has been detected. Training data and testing are independent from each other and test data are unseen by the model which increases the difficulty.

This technique verifies or identifies a person using a detected face in the previous stage. For this, there is a need to use various algorithms like HOG, COTS, etc. After applying all these algorithms, their performances are compared, to check which algorithm is best suited for data set. Then training and testing the model is done. Several methods can be used to check the performance of a model like through confusion matrix.

4.4   EMPIRICAL EVALUATION

4.4.1   DATA PRE-PROCESSING AND BALANCING OF DATA

The images and videos which had noise and had low resolution were removed. Then normalize all the face images to identical lighting conditions. We reduce resolution variation in different videos to increase the final accuracy. Face components like eyes, nose, mouth, etc., are extracted for detection and recognition. It is important to extract the features for initial processes like face tracking and recognition. This meaningful data extracted from face image will help to detect a person.

4.4.2   SPLITTING OF THE DATA SET

The data set is divided into two parts where 30% data as a testing data set and the rest is considered as training data. Training data and testing are independent from each other and test data is unseen by the model which increases the difficulty.

4.4.3   CHOOSING THE MODEL

Two algorithms have been used to detect faces which are viola jones and tiny faces to cover all the lighting conditions. These two approaches cover every situation of data set which helps in increasing the accuracy.

4.4.4   FACE DETECTION TECHNIQUES

In the proposed approach, we first computed a baseline for person detection with the help of two face detectors. These detectors are named as Viola Jones and Tiny faces. Both models demonstrate recall and precision values with respect to efficiency. For the detection of faces in crowds with lower resolution, here the model named Tiny faces is already pre-trained.

4.4.5   VIOLA JONES

Viola Jones is used here to meet the competitive world with a sufficient rate of detection. The main purpose of Viola Jones is to detect faces in a particular frame of video. For humans this task is quite easy but if the same task is to be performed in computers then there are some protocols to be followed. Full view of the upright front face is required by the Viola Jones model to perform well in a managed manner. This model needs the target faces to be pointed towards the UAVs camera device with minimal possible tilt in any direction. Also, it is to be noted that this restriction can lead it to less reliability and less functionally in real world problems because in general recognition is performed after detection. On the other hand, despite all these constraints, this algorithm can perform fast detection of faces in comparison to other algorithms. Accuracy of detected faces in each iteration of Viola Jones is shown in Table 4.1.

Viola Jones needs a proper frontal view of the face, and in the above input, two faces are clearly visible. This is the reason that for these types of images and screenshots, Viola Jones is able to give good accuracy and recall. For the images in which faces are not clear, it will not be able to detect the faces accurately. Tiny Faces: Detecting faces in crowded areas has certain challenges like the captured footage gives small faces if shot from distance to cover an area and also some details are also missed as images are of low resolution. Input to Viola Jones model and its output are shown in Figures 4.3 and 4.4.

It is the result of the tiny faces algorithm. The size of the objects is small; still this approach is able to detect it which increases the performance. Because of this advantage, tiny faces give better accuracy than viola jones. Face Recognition Techniques: After face detection, face recognition is done. This technique verifies or identifies a person using a detected face in the previous stage. Now need to use various algorithms like HOG, COTS, etc. After applying all these algorithms, there is need to compare their performance to check which algorithm is best suited for data set.

TABLE 4.1

Represents the Accuracy of Detected Faces in Each Iteration of Viola Jones

Number of Detections: 2	Accuracy (%)
Person	96.6
Person	92.0

FIGURE 4.3　Input to Viola Jones model.

FIGURE 4.4   Output of LBP.

Then training and testing the model is done. Input and output for tiny faces model are shown in Figures 4.5 and 4.6.

HOGs: The HOG can be a function identifier used for image recognition purposes in computer vision applications. It uses a frame for sliding detection that travels around images and then categorizes them. By allocating the strength variation or surface orientation, the HOG descriptor aims to clarify the framework of the subject. The image is split into multiple connected regions known as cells, and in each cell a histogram of gradient directions is compiled for pixels. Mixing these histograms produces a descriptor. That is why the HOG descriptor is appropriate for the identification of faces. Table 4.2 represents the accuracy of detected faces in each iteration of Tiny Faces model. Table 4.3 represents the accuracy of recognized faces in each iteration of HOG model. Figures 4.7 and 4.8 show input and output of HOG images.

FIGURE 4.5   Input to Tiny Faces Model.

FIGURE 4.6   Output of Tiny Faces Model.

TABLE 4.2
Represents the Accuracy of Detected Faces in Each Iteration of Tiny Faces Model

S.No.	Number of Detections:13	Accuracy (%)
1	Person	91.7
2	Person	77.9
3	Person	63.4
4	Person	58.9
5	Person	71.2
6	Person	78.7
7	Person	59.7
8	Person	50.3
9	Person	89.0
10	Person	74.8
11	Person	78.9
12	Person	98.9
13	Person	89.4

TABLE 4.3
Represents the Accuracy of Recognized Faces in Each Iteration of HOG Model

Number of Recognitions:9	Accuracy (%)
Face1	88.9
Face2	97.9
Face3	73.4
Face4	58.9
Face5	71.2
Face6	78.7
Face7	59.9
Face8	70.3
Face9	89.0

FIGURE 4.7 Input to HOG.

FIGURE 4.8   Output of HOG.

4.4.6   LOCAL BINARY PATTERN

LBP could be a simple, but particularly significant, experience role conflict which marks the components of an image by convolving each sensor's neighborhood and treats the effect as a probability sample. The extracted feature controller seems to have become a correct strategy in diverse products due to its exclusionary energy and computer elegance. The solution to the traditionally disparate conceptual and systemic frameworks of image segmentation should be seen as a unitary one. Perhaps the LBP driver's most attractive characteristic in real-world implementations is its resistance to monotonous Gray-scale shifts triggered, for example, through fluctuations in lighting. Its device usability is yet another important component, which makes it possible to analyze images in lot of difficulty span environments. Accuracy of recognized faces in each iteration of LBP model and the accuracy of recognized faces in each iteration of VGG model are shown in Tables 4.4 and 4.5, respectively. Input and output for LBP faces are depicted in Figures 4.9 and 4.10.

Unlike Haralick texture options that calculate a worldwide illustration of texture supported by the gray level co-occurrence matrix, LBPs instead calculate an area illustration of texture. This native illustration is made by comparing every constituent with its close neighborhood of pixels. It is a deep learning primarily based feature extractor, which labels the constituent of a picture. VGG-Face: Visual Geometry Group-Face encodes the face to represent 2048 numbers. Then it measures the Euclidean distance between two encoded faces. If it is the same face then the value

---

TABLE 4.4

Represents the Accuracy of Recognized Faces in Each Iteration of LBP Model

Number of Recognitions: 9	Accuracy (%)
Face1	71.2
Face2	58.9
Face3	77.9
Face4	47.3
Face5	71.2
Face6	78.7
Face7	59.7
Face8	58.9

TABLE 4.5

**Represents the Accuracy of Recognized Faces in Each
Iteration of VGG Model**

Number of Recognitions: 9	Accuracy (%)
Face1	59.7
Face2	74.8
Face3	63.4
Face4	58.9
Face5	71.2
Face6	91.7
Face7	59.7
Face8	50.3
Face9	89.0
Face10	74.8
Face11	78.9

FIGURE 4.9   Input of Local Binary Pattern (LBP).

FIGURE 4.10   Output of LBP.

is low, otherwise it is high. The VGG design uses a profoundly convolutional neural network architecture, and convolutional layers with kernels and ReLU activation functions.

4.4.7   COMMERCIAL-OFF-THE-SHELF SYSTEM

By using specular reflection feature, blurriness feature, chromatic moment feature, color diversity feature, COTS does image distortion face detection. But it cannot

FIGURE 4.11   Input to VGG-Face.

FIGURE 4.12   Output of VGG-Face.

differentiate between genuine and spoof faces. Since COTS uses an integrated face detector, it results in a high failure to detect sample levels, resulting in lower efficiency. Applying model on Real Test Data and Generating the output, several methods can be used to check the performance of a model like through confusion matrix by calculating false positive, false negative, true positive and true negative. We can also use F1 score, mean absolute error, etc. In this project, we have measured the performance by giving scores to each box that has been detected. Accuracy of Recognized Faces in Each Iteration of COTS Model is presented in Table 4.6. Input and output of VGC face are shown in Figures 4.11 and 4.12.

Performance Measures:

TABLE 4.6

Represents the Accuracy of Recognized Faces in Each Iteration of COTS Model

Number of Recognitions: 9	Accuracy (%)
Face1	81.7
Face2	78.4
Face3	83.2
Face4	58.9
Face5	41.2
Face6	78.7

FIGURE 4.13   Input to Commercial-Off-The-Shelf system (COTS).

FIGURE 4.14   Output of COTS.

4.4.8   FACE DETECTION

Viola Jones does not give good precision and recall because it needs a proper frontal view of the face. Data set contains the videos of crowded places also for which Viola Jones does not perform well.

Table 4.7 represents the precision and recall for viola jones and tiny faces. According to the result, tiny faces give better precision and recall as compared to viola jones. This result is justifiable because the tiny face detector is trained for low resolution data which increases its performance.

4.4.9   FACE RECOGNITION

In framewise identification results, it can be observed that for all frames, VGG-Face algorithm gave the best results with 14.36% in identification performance. The LBP model gave the accuracy of 5.08%, on the other hand, 5.08% accuracy was achieved with the HOG algorithm. The COTS algorithm gave the worst results in

TABLE 4.7

Represents the Comparison between Face Detection Models

Algorithm	Precision	Recall
Viola Jones	20.01	25.50
Tiny Faces	91.83	90.02

TABLE 4.8
Represents the Comparison between Face Recognition Models

Algorithm	Accuracy
HOG	6.66
LBP	4.26
VGG-Face	14.36
COTS	3.26

comparison to others by giving the accuracy of less than 7%. Comparison between Face Recognition Models are shown in Table 4.8.

4.5 CONCLUSION AND FUTURE WORK

In this chapter, the main focus is on increasing the performance of existing models for detecting and recognizing the faces in drone shot videos after overcoming the challenges faced. It was hard to detect faces in crowded places and the accuracy was low. To overcome this challenge, the tiny faces algorithm is used that is able to detect small objects which in turn lead to increased performance. As Viola Jones needs a proper frontal view of the face, this algorithm is unable to detect faces in crowded places. According to the result, tiny faces give better precision and recall as compared to Viola Jones because tiny face detectors are trained for low resolution data. In framewise identification results, it can be observed that for all frames, the VGG-Face algorithm gave the best results and COTS gave the worst accuracy for face recognition in drone shot videos.

In the end, the expected accuracy and outcomes were received and by using these algorithms. For future work, there are some challenges that are yet to be solved. One such challenge is the variations in the poses of different faces which reduces the performance and accuracy. Further work can be done for more improvement. Current face recognition techniques do not perform well in data set. Due to which, it faces some challenges which reduce the performance. The distance between the object and the drone is too large in some cases, which makes it hard for the model to give accurate results. Variation in poses also reduces the performance of face detection. There are some approaches can use which may handle the above challenges but then it will be hard for the model to distinguish between objects.

REFERENCES

1. Schumann, Arne, and Tobias Schuchert. "Person re-identification in UAV videos using relevant feedback." *Video Surveillance and Transportation Imaging Applications* 2015. Vol. 9407. International Society for Optics and Photonics, 2015.
2. Passalis, Nikolaos, and Anastasios Tefas. "Concept detection and face pose estimation using lightweight convolutional neural networks for steering drone video shooting." *2017 25th European Signal Processing Conference (EUSIPCO).* IEEE, 2017.
3. Grigorev, Aleksei, et al. "Deep person re-identification in UAV images." *EURASIP Journal on Advances in Signal Processing,* 2019(1), 54, 2019.

4. Layne, Ryan, Timothy M. Hospedales, and Shaogang Gong. "Investigating open-world person re-identification using a drone." *European Conference on Computer Vision.* Springer, Cham, 2014.

5. Mademlis, Ioannis, et al. "Challenges in autonomous UAV cinematography: An overview." *2018 IEEE International Conference on Multimedia and Expo (ICME).* IEEE, 2018.

6. M. Bindemann, M. C. Fysh, S. S. Sage, K. Douglas, and H. M. Tummon. Person identification from aerial footage by a remote-controlled drone. *Scientific Reports*, 7(1), 2017.

7. Q. Cao, L. Shen, W. Xie, O. M. Parkhi, and A. Zisserman. Vggface2: A dataset for recognising faces across pose and age. In *IEEE FG*, pages 67–74, 2018.

8. H.-J. Hsu and K.-T. Chen. DroneFace: An open dataset for drone research. *ACM MMSys*, pages 187–192, 2017.

9. I. Kalra, M. Singh, S. Nagpal, R. Singh, M. Vatsa, and P. Sujit. Drone surf: Benchmark dataset for drone-based face recognition. *IEEE International Conference on Automatic Face and Gesture Recognition*, 2019.

10. P. Zhu, L. Wen, X. Bian, L. Haibin, and Q. Hu. Vision meets drones: A challenge. *arXiv preprint arXiv:1804.07437*, 2018.

11. M. Singh, R. Singh, and A. Ross. A comprehensive overview of biometric fusion. *Information Fusion*, 2019.

12. M. Khari, R. Dalal, and P. Rohilla. Extended Paradigms for Botnets with WoT Applications: A Review. *Smart Innovation of Web of Things*, 105, 2020.

13. M. Khari, R. Dalal, U. Misra, and A. Kumar. AndroSet: An automated tool to create datasets for android malware detection and functioning with WoT. *Smart Innovation of Web of Things*, 187, 2020.

5 ECG-Based Biometric Authentication Systems Using Artificial Intelligence Methods

Sanjeev Kumar Saini
ABES Engineering College

Guru Gobind Singh
Indraprastha University

Rashmi Gupta
NSUT East

CONTENTS

DOI: 10.1201/9781003138068-5

5.1   INTRODUCTION

Individual identification is a global need nowadays in various fields like banking, buildings, airports, defense, healthcare, aviation, law enforcement, attendance management, etc. In past decades the personal identification has been performed on various platforms like smart phones, computers, building security, etc., through different techniques including personal identification number (PIN), password, and radio frequency-based ID (RFID) cards. These identification methods are categorized as knowledge based identification systems including passwords, PINs, etc. and token based an identification system which includes ID cards, software key, etc. These traditional methods of personal identification are not secure and have limitations of being theft, lost, and misuse [1]. In contrast to traditional methods, biometric-based identification has emerged as an effective and secure way of individual authentication in recent years [2,3]. The biometric characteristics have advantages of uniqueness, measurable, and permanent that makes the biometric identification systems robust and secure from fraud, theft, and misuse [4].

5.2   BIOMETRIC IDENTIFICATION

Biometrics ('bio' means biology and 'metrics' means measurement) refers to the measurement and analysis of biological signals acquired human body reflecting the physiological or behavioral traits of a person. Biometric identification is an automatic technique to use physiological or behavioral characteristics of a person to authenticate his/her identity [3]. In past, physiological characteristics including finger print, facial geometry, palm geometry, iris pattern, finger vein, DNA, ECG, etc., and behavioral characteristics including voice, signature, gait, keystroke, etc. have been used to develop biometric identification systems [5,6]. Each of these biometric identifiers is universal and distinct in humans. In contrast to traditional knowledge-based identifiers, biometric identifiers have advantages like universality, uniqueness, measurability, and robustness. The biometric identifiers commonly used in a biometric identification system are shown in Figure 5.1.

Over the years biometric identification devices have been developed using one or more physiological and/or behavioral characteristics as biometric identifier. Each of the identifiers has some advantages and limitations. With the advancements in sensor technology and signal/image processing methods, the new trend in human identification is multimodal biometric system that uses more than one physiological characteristic (finger print, face, iris, vein, voice, etc.) of a person to achieve a more secure and reliable authentication system. Figure 5.2 shows the trends and future prospects for different biometric identifiers [7].

5.2.1   BIOMETRIC IDENTIFICATION SYSTEM ARCHITECTURE

A biometric identification system is developed in two modules, namely, enrollment module and identification module as shown in Figure 5.3. The enrollment phase is a learning phase during which a biometric trait of a person is captured through a biometric sensor, processed through a, and passed through a feature extraction stage.

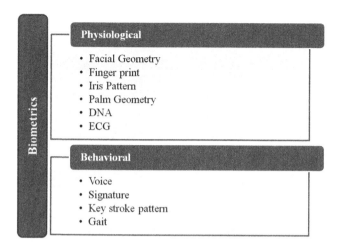

FIGURE 5.1   Biometric identifiers used in a biometric identification system.

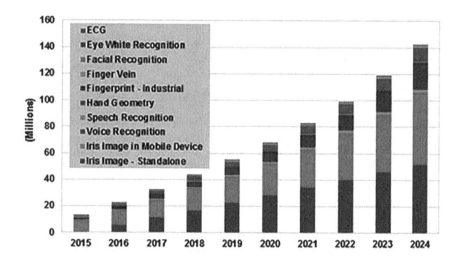

FIGURE 5.2   Trends and future prospects of biometric identifiers.

The extracted features are then stored in a central database in a suitable digital form as a sample characteristic, called as 'template'. During the identification mode the system captures the same biometric trait of a person and same features are extracted in digital form. The resulting digital sample characteristic is then presented to a template matching stage to establish a match with the templates stored in database to authenticate a person's identity.

The performance of a biometric identification system is evaluated in terms of identification accuracy, response time, storage capacity, and robustness. The identification accuracy is a major concern for a biometric identification system which

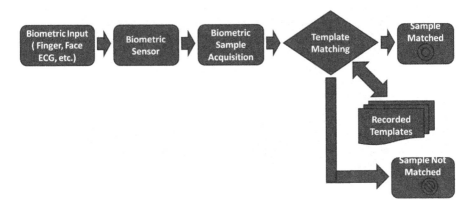

FIGURE 5.3   Biometric identification system architecture.

is measured using false recognition rate (FRR) and false acceptance rate (FAR) parameters. FRR is a measure of probability of false rejection of a correct biometric input by the system and FAR is a measure of probability of false acceptance of a wrong biometric input by the system. Biometric identification devices have vast application areas including attendance management, airport, healthcare, personal and commercial building security, aviation, financial transactions, smart phone and personal computer security, etc.

5.3   ECG-BASED BIOMETRIC IDENTIFICATION

Electrocardiogram (ECG) has been a popular biological signal for researchers in various fields to develop automatic healthcare systems including cardiac arrhythmia detection, emotion detection, etc. [8,9]. In recent years, apart from healthcare systems, ECG signals are emerged as a reliable biometric identifier to develop standalone or multimodal biometric human authentication devices [6,10]. Researchers have used different features (inter pulse interval (IPI), heart rate variability (HRV), etc.) extracted from ECG signals as a unique identifier for biometric identification. Before using the ECG signal as a biometric identifier several steps of preprocessing, segmentation, feature extraction, and feature selection are required. In contrast to other biometric identifiers like finger print, iris scan, hand geometry, etc., ECG as a biometric identifier has advantage that it can be captured from different parts of body including finger, toes, wrist, etc.

Similar to other biometric identification systems, ECG-based biometric identification systems are pattern recognitions systems which perform template matching to match a template ECG stored in central database with the presented ECG within a predefined confidence level. For a given threshold value of confidence level, the system produces a matched or unmatched result. With the advancements in artificial intelligence and ML approaches, biometric authentication systems have been developed by researchers with high level of accuracy and recognition rate. Figure 5.4 describes the general architecture of an ECG-based biometric identification system.

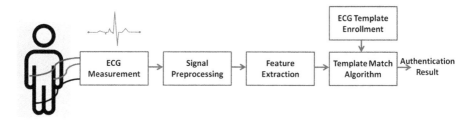

FIGURE 5.4 Architecture of ECG-based biometric identification system.

5.3.1 ECG PHYSIOLOGY

Figure 5.5 shows the anatomy of human heart. The human heart contains four chambers (atria and ventricles) of heart muscles which contracts and expand in a rhythm to supply the oxygen rich blood through lungs to all parts of body. The heart cells and surrounding fluid consist of electrolytes (sodium, potassium, and calcium, etc.) and the interchange of these electrolytes between cell body and external fluid gives rise to electrical activity in the heart cells [11].

The mechanical expansion and relaxation of heart muscles is carried out by the electrical impulses that are stimulated from sinoatrial node (SA node) or sinus node and spread across heart chambers through atrioventricular node (AV node), bundle of His, and purkinje fibres (PF). This electrical activity in the heart results in action potentials consisting of depolarization (contraction of atria or ventricle) and repolarization (relaxation of atria or ventricle) stages [12] as shown in Figure 5.6.

5.3.2 ECG WAVEFORM

ECG is a record of electrical potentials generated in heart during the cardiac cycle [13]. These bio potentials can be acquired from an array of skin electrodes placed at different positions on human body and plotted as a waveform over time axis. The ECG waveform contains a sequence of events with standard amplitude and duration corresponding

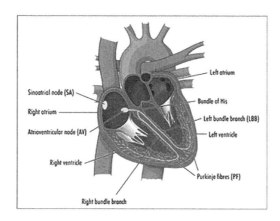

FIGURE 5.5 Anatomy of human heart.

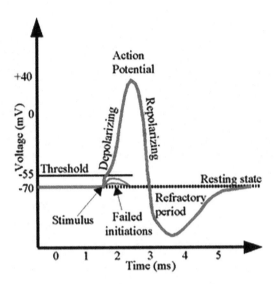

FIGURE 5.6   Action potential during cardiac cycle.

to the depolarization and repolarization activity of heart [14]. The important parts of waveform that contains important information and completely characterizes the heart operation are P wave, QRS complex, and T wave as shown in Figure 5.7.

Different activities during cardiac cycle and associated ECG segment are as follows:

- The P wave corresponds to the depolarization that originates from SA node and spreads throughout atria. The PR interval represents the time duration between the onset of atrial depolarization and ventricular depolarization which ranges from 0.12 to 0.2 s for a normal heart.
- The QRS segment of ECG waveform corresponds to the ventricular depolarization which starts at the end of P wave. The duration of ventricular depolarization ranges from 0.06 to 0.1 s for a normal heart. The shape of QRS complex may vary depending on the electrode position during ECG recording.
- The T wave is a positive deflection that corresponds to the ventricular repolarization of the heart muscles.

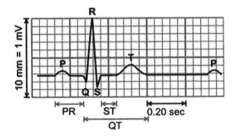

FIGURE 5.7   Important segments of ECG waveform.

TABLE 5.1
ECG Segments and Their Standard Values

ECG Segment	Amplitude	ECG Segment	Duration
P	0.25 mV	P	0.08–0.11 s
Q	25 % of R wave	PR	0.12–0.2 s
R	1.6 mV	QT	0.35–0.44 s
T	0.1–0.5 mV	ST	–0.15 s

The standard values of different segments of the ECG waveform for a normal functioning heart are listed in Table 5.1 [15].

5.4 ECG SIGNAL PROCESSING FOR BIOMETRIC SYSTEMS

ECG signals are recorded using automated digital equipment to measure the amplitudes and durations during cardiac activity as electronic potentials captured by skin electrodes placed at different positions. The morphology of the recorded ECG signals depends on the placement of electrodes in reference to a lead system (for example 12-lead system) used for the recording [16]. The recorded ECG signals are weak (amplitude in mV) and distorted signals that are not suitable for further analysis. Before the ECG signal can be applied to a classification system it is to be processed through several stages. The general steps required for ECG signal processing are shown in Figure 5.8.

5.4.1 DENOISING

During the recording process ECG signals are mixed with distortion or noise due to several internal and external factors. Internal factors include prolonged depolarization etc. and external factors include power line interference, base line wander, muscle movements, respiration, and electrode movements [17,18]. Before using the raw ECG signals for biometric authentication, suitable preprocessing techniques should be applied to improve the signal-to-noise ratio and extract the correct features from the ECG signals. Different filtering techniques including Kalman filtering, adaptive filtering, fixed notch filtering, median filtering, and frequency domain analysis techniques like fast Fourier transform (FFT) and discrete wavelet transform (DWT) have been used.

5.4.2 SEGMENTATION

After the noise removal the continuous ECG signal is segmented into individual heartbeats to extract the information about start point, peak amplitude, and end points from P, QRS, and T waves [19]. R peak in QRS complex is considered as a reference point during the segmentation process of ECG waveform. A segmentation algorithm locates the R-peak in an ECG signal and remaining points of interest are then located by sliding a window of fixed duration around the located R-peak.

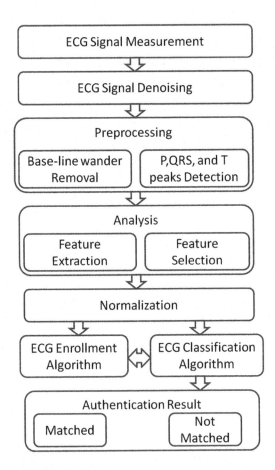

FIGURE 5.8   Steps in ECG signal processing.

Several methods for QRS detection and ECG segmentation including Pan-Tompkins algorithm [20], autoregressive model [21], time warping [22], deep learning [23] have been used. Figure 5.9 shows an ECG signal after denoising and a heartbeat segment [24].

5.4.3   FEATURE EXTRACTION

Feature extraction is a method to detect the important characteristics from a given signal. Using a suitable feature extraction method, the given signal is represented as a set of features which is further used to train a mathematical model [25]. Various methods including autoregression (AR), fast Fourier transformation (FFT), wavelet transformation (WT), linear prediction (LP), independent component analysis (ICA) are available to extract the ECG features in time-domain, frequency-domain, and joint time-frequency domain [26]. Time-domain features are simple and provide temporal information for peaks and durations of events in the ECG signal, while frequency-domain features provide accurate information for

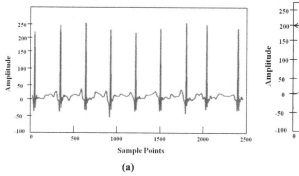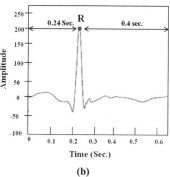

FIGURE 5.9   (a) A denoised ECG signal. (b) Segmented heartbeat.

different frequency components in the signal. As ECG signals are non-stationary in nature, time-frequency domain features extracted using wavelet transform provide multi-resolution analysis of the ECG signals and have been widely used in literature [27–30]. Different methods used for feature extraction in time, frequency, and time-frequency domains are listed in Table 5.2.

5.4.4   FEATURE SELECTION

The selection of appropriate set of optimal features from the extracted features has a great impact on model training and design of biometric authentication systems [31]. Feature selection is important to find out the dominant features and remove the redundant, noisy, and unwanted features from the original features to achieve the better performance of the classification model. Among the methods like principal component analysis (PCA), Best First Search (BFS), wavelet decomposition, and non-overlap area distribution measurement, genetic algorithm (GA) has been widely used for optimal feature selection with good accuracy [32].

TABLE 5.2
ECG Feature Extraction Methods

Feature Extraction Domain	Extraction Methods	Applications
Time-domain	• Autoregressive modeling • Linear predictive coding • Kernel based modeling	Emotion recognition, mental stress analysis, biometric authentication, etc.
Frequency-domain	• Fast Fourier transform • Discrete cosine transform • Hilbert transform	ECG signal denoising, signal compression, arrhythmia detection, etc.
Time-frequency domain	• Short-term Fourier transform • Continuous wavelet transform • Discrete wavelet transform	Cardiac arrhythmia recognition, biometric authentication, medical image classification, etc.

An optimal feature set was selected using GA and selected features were used to classify the heart beats using a decision tree (DT)-based classification model with accuracy of 86.96% [33].

5.5 ECG-BASED BIOMETRIC AUTHENTICATION SYSTEMS

In recent years development of biometric authentication systems using artificial intelligence and ML techniques has been a topic of research. Several methods have been developed in literature for automatic biometric identification using live data from human body with state-of-the-art ML techniques including neural networks, fuzzy logic, etc. [34–40]. Among these biometric identification systems, ECG or EKG has emerged as a new biometric modality to be used as a biometric identifier. However, the use of ECG as a biometric identifier was introduced in 1977 in US military report [35], still there are challenges of live data acquisition through wearable devices, pre-processing, and selection of appropriate features from ECG recordings. Figure 5.10 demonstrates the general framework for an ECG-based biometric authentication system using ML technique.

A new method for human identification was proposed by Biel et al. in 2001 [36]. The 12-lead ECG was recorded from 20 persons in rest position; 30 features were extracted from the ECG recordings and classification was done using soft independent modeling of class analogy (SIMCA) to identify a person from the recorded database.

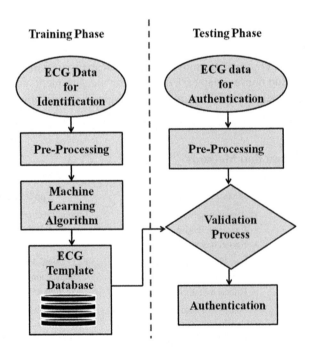

FIGURE 5.10   General scheme for ECG biometric authentication using ML.

5.6   ARTIFICIAL INTELLIGENCE METHODS FOR ECG-BASED BIOMETRIC AUTHENTICATION SYSTEMS

An ECG-based biometric authentication technique based on supervised fuzzy neural network model was developed using PhysioNet database containing 73 ECG recordings [37]. After noise removal from raw ECG signal, Haar wavelet transformation was performed to extract 29 wavelet features of ECG signal based on a4, d3, and d4 coefficients. The selected features were utilized to train (biometric enrollment) a neural network with weighted membership function (WMF), which consists of an input layer, hyperbox layer, and output layer. The number of nodes in the input layer corresponds to the number of ECG features and the number of nodes in hyperbox layer is computed as number of nodes in input layer times the number of nodes in output layer (two nodes in case of biometric authentication). The nodes in hyperbox layer are grouped, each group containing number of nodes equal to number of nodes in input layer, and each group is mapped to each ECG feature connected to input layer. Each node in a hyperbox group is associated with a WMF that generates a boundary sum of WMFs to classify the input features into classes at output layer. Figure 5.11 shows the structure of a fuzzy neural network model with WMFs at hyperbox layer with input layer containing five nodes and output layer containing two nodes. The model was tested on 73 ECG recordings from PhysioNet database with true acceptance rate 98.32% and FAR 5.84%.

A biometric authentication system suitable for security checks and hospitals was described using ECG as a biometric identifier [35]. The time-sliced ECG data was used to train and test a ML model. The performance of DT and support vector machine (SVM) models for biometric authentication was evaluated using PhysioNet ECG database after preprocessing and feature selection based on mutual information theory [38]. DT regression method was selected after comparing the root mean square error (RMSE), mean absolute error, and training time for DT and SVM training for ECG time sliced data. An overall accuracy of 92.7% using DT model for the authentication system was achieved.

The design of a sparse autoencoder neural network was suggested for ECG-based biometric authentication system [24]. An autoencoder neural network is a unsupervised learning technique based on back propagation in which the target output $(\hat{x})$

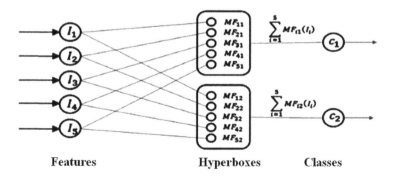

FIGURE 5.11   Fuzzy neural network model with WMFs.

is set equal to input (x). The autoencoder then learns a mapping function such as to minimize the error between target and input [39]. Figure 5.12 shows the structure of an autoencoder with input, hidden, and output layer. The encoder stage maps the input data into a middle representation $(\hat{x})$ through a nonlinear activation function (sigmoid or other) and decoder stage aims to reconstruct the original input data with minimum reconstruction error and thus finds a function $h_{w,b}(x) \approx x$; w is the weight matrix and b is bias vector of network. The encoding and decoding operations are explained in following equations:

The encoding transforms the input x into middle representation $\tilde{x}$ using equations (5.1) and (5.2).

$$\tilde{x} = h(x) = f\left(\sum_{j=1}^{m} w_1 . x + b_1\right) \tag{5.1}$$

$$f(z) = 1 / (1 + \exp(-z)) \tag{5.2}$$

The decoding reconstructs the output y from middle representation $\tilde{x}$ using equation (5.3).

$$y = f(w_2 . \tilde{x} + b_2) \tag{5.3}$$

Using the backpropagation algorithm the model trains continuously by adjusting the weight matrix and bias vector to minimize the reconstruction error. To further fine-tune the learned features, softmax classifier is added next to the autoencoder that performs the back propagation algorithm for feature classification. After the model is trained, test data were applied to trained model to classify the matched or unmatched feature at the output softmax layer. The model was evaluated on MIT-BIH-AHA

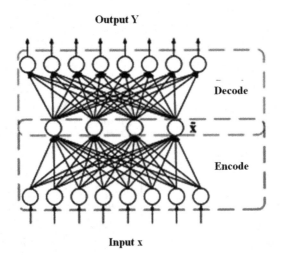

FIGURE 5.12 Autoencoder neural network structure.

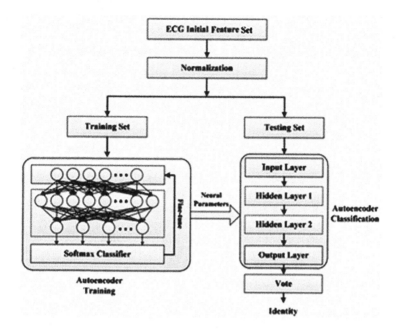

FIGURE 5.13 Flow chart of autoencoder-based biometric authentication model.

ECG database with accuracy of 96.82%. The flow chart for the biometric authentication model is shown in Figure 5.13.

A simple and fast algorithm for ECG biometric authentication was proposed in [40] using heart beat as a biometric identifier. In the first-stage Hamilton's method [41] wad was used to detect the heart beat in a short-term ECG signal and segmentation was performed on the detected heart beat taking R-peak as a reference point for segmentation. In the second stage of classification a residual depth-wise convolutional neural network (CNN) was utilized to classify the heart beats. The model was validated on MIT-BIH ECG database with an authentication accuracy of 97.92%. In a work by Belo et al. [42], the performance of two deep learning models, CNN, and recurrent neural network (RNN) was compared for an ECG biometric authentication system. Both models were tested on MIT-BIH ECG database and authentication accuracies of 92.7% and 96.3% for recurrent neural network and CNN, respectively.

5.7 CONCLUSION

Biometric authentication using ECG signals has been a topic of research during last two decades and emerged as a reliable biometric authentication technique but still there are challenges for capturing ECG sample using simple wearable devices, signal processing, and simple but fast classification algorithms. Over the traditional knowledge-based (PIN, password, etc.) and token-based (RFID, bar code, etc.) person identification systems, ECG-based authentication systems are becoming popular as strong and reliable authentication systems as ECG as a biometric identifier

TABLE 5.3

Performance Comparison of ECG-Based Biometric Authentication Systems

Reference No.	Database	Biometric Features	Classification Method	Authentication Accuracy
[24]	MIT-BIH-AHA database	Time-frequency domain features derived from discrete wavelet transform (DWT) of QRS segments.	Sparse autoencoder Neural Network (NN)	96.82%
[35]	Physio Bank database	Time-sliced ECG signal using multi variable regression with R-peak as reference	DT and SVM models	92.77%
[36]	12-lead recorded ECG signal in rest position through experiment on 20 persons	30 temporal features selected using correlation matrix method.	SIMCA	98%
[37]	PhysioNet database	Wavelet coefficients(a4, db3, and db4) extracted using Haar wavelet transformation	Neural Network (NN) with fuzzy membership function	98.32%
[40]	MIT-BIH database	R-peak positions derived from Hamilton's method of ECG segmentation	Residual Depthwise CNN	97.92%
[42]	MIT-BIH database	Denoised ECG signals	CNN and RNN	96.3% (CNN) 92.7 % (RNN)

provides the features of uniqueness, universal, and free from theft and misuse. With the recent developments in the area of ML and deep learning techniques new opportunities are available to develop a simple, fast, and accurate ECG-based biometric authentication system.

In this chapter, we explored the use of ECG signals as biometric identifiers for the development of biometric authentication systems. The origin and physiology of ECG signals are discussed considering their important characteristics useful for biometric identifier. Different methods for signal processing steps required for generating the feature set and identification template are explained. Various ML and artificial intelligence methods used by the researchers to develop ECG-based biometric authentication systems are compared in Table 5.3. Accuracy of different artificial intelligence methods for ECG-based biometric authentication is shown in Figure 5.14.

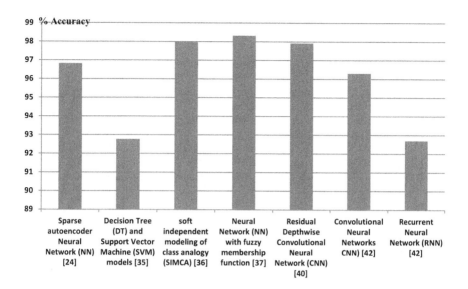

FIGURE 5.14 Comparative performance of classification methods for biometric authentication.

REFERENCES

1. L. Hong, S. Pankanti, and N. J. Hawthorne, "Biometrics: Promising frontiers for emerging identification market," *Commun. ACM - CACM*, 2000.
2. M. Dua, R. Gupta, M. Khari, and R. G. Crespo, "Biometric iris recognition using radial basis function neural network," *Soft Comput.*, vol. 23, no. 22, pp. 11801–11815, 2019.
3. M. N. Dar, M. U. Akram, A. Shaukat, and M. A. Khan, "ECG based biometric identification for population with normal and cardiac anomalies using hybrid HRV and DWT features," *2015 5th Int. Conf. IT Converg. Secur. ICITCS 2015- Proc.*, 2015.
4. A. M. and A. A. Souhail Guennouni, "Biometric Systems and Their Applications," Intech, no. tourism, p. 13, 2016.
5. S. Hadiyoso, I. Wijayanto, and E. M. Dewi, "ECG Based Biometric Identification System using EEMD, VMD and Renyi Entropy," *2020 8th Int. Conf. Inf. Commun. Technol. ICoICT 2020*, pp. 15–19, 2020.
6. F. Sufi, I. Khalil, and J. Hu, "ECG-Based Authentication," *Handb. Inf. Commun. Secur.*, pp. 309–331, 2010.
7. "Global Biometric Market Analysis: Trends and Future Prospects." [Online]. Available: https://www.bayometric.com/global-biometric-market-analysis/.
8. AlGhatrif Majd and Lindsay Joseph, "A brief review: history to understand fundamentals of electrocardiography," *J. Community Hosp. Intern. Med. Perspect. 2012.*, vol. 1, pp. 301–332, 2012.
9. S. Kaplan Berkaya, A. K. Uysal, E. Sora Gunal, S. Ergin, S. Gunal, and M. B. Gulmezoglu, "A survey on ECG analysis," *Biomed. Signal Process. Control*, vol. 43, pp. 216–235, 2018.
10. S. A. El_Rahman, "Biometric human recognition system based on ECG," *Multimed. Tools Appl.*, vol. 78, no. 13, pp. 17555–17572, 2019.
11. M. Sampson and A. McGrath, "Understanding the ECG. Part 1: Anatomy and physiology," *Br. J. Card. Nurs.*, vol. 10, no. 11, pp. 548–554, 2015.
12. R. G. Carroll, "The Heart," in *Elsevier's Integrated Physiology*, Elsevier, 2007, pp. 65–75.

13. S. K. Saini and R. Gupta, "A review on ECG signal analysis for mental stress assessment," *Proc. 2019 6th Int. Conf. Comput. Sustain. Glob. Dev. INDIACom 2019*, pp. 915–918, 2019.
14. R. E. Klabunde, *Cardiovascular Physiology Concepts*, Second edition. Lippincott Williams & Wilkins, 2012.
15. M. S. Al-Ani, "ECG waveform classification based on P-QRS-T wave recognition," *UHD J. Sci. Technol.*, vol. 2, no. 2, p. 7, 2018.
16. L. Freudzon *et al.*, "Electrocardiographic monitoring," *Kaplan's Essentials Card. Anesth. Card. Surg.*, pp. 168–202, 2017.
17. V. Gupta and M. Mittal, "A comparison of ECG signal pre-processing using FrFT, FrWT and IPCA for improved analysis," *Irbm*, vol. 40, no. 3, pp. 145–156, 2019.
18. A. Merrikhi, H. R. Asadabadi, A. A. Beigi, S. M. Marashi, H. Ghaheri, and Z. N. Zarch, "Electrocardiogram (ECG) signal processing," *Med. J. Islam. Repub. Iran*, vol. 28, no. 1, pp. 1–16, 2014.
19. I. Beraza and I. Romero, "Comparative study of algorithms for ECG segmentation," *Biomed. Signal Process. Control*, vol. 34, pp. 166–173, 2017.
20. J. Pan and W. J. Tompkins, "A real-time QRS detection algorithm," *IEEE Trans. Biomed. Eng.*, vol. 32, no. 3, pp. 230–236, 1985.
21. L. Szilágyi, S. M. Szilágyi, A. Frigy, L. Dávid, and Z. Benyó, "Quick ECG segmentation, artifact detection and risk estimation methods for on-line Holter monitoring systems," *IFMBE Proc.*, vol. 14, no. 1, pp. 1021–1025, 2007.
22. H. J. L. M. Vullings, M. H. G. Verhaegen, and H. B. Verbruggen, "ECG segmentation using time-warping," *Lect. Notes Comput. Sci. (including Subser. Lect. Notes Artif. Intell. Lect. Notes Bioinformatics)*, vol. 1280, pp. 275–285, 1997.
23. V. Moskalenko, N. Zolotykh, and G. Osipov, "Deep learning for ECG segmentation," *Stud. Comput. Intell.*, vol. 856, pp. 246–254, 2020.
24. D. Wang, Y. Si, W. Yang, G. Zhang, and J. Li, "A novel electrocardiogram biometric identification method based on temporal-frequency autoencoding," *Electron.*, vol. 8, no. 6, pp. 1–24, 2019.
25. S. Krishnan and Y. Athavale, "Trends in biomedical signal feature extraction," *Biomed. Signal Process. Control*, vol. 43, pp. 41–63, 2018.
26. F. M. Vaneghi, M. Oladazimi, F. Shiman, A. Kordi, M. J. Safari, and F. Ibrahim, "A comparative approach to ECG feature extraction methods," *Proc. -3rd Int. Conf. Intell. Syst. Model. Simulation, ISMS 2012*, pp. 252–256, 2012.
27. R. F. Hassan and S. A. Shaker, "ECG signal de-noising and feature extraction using discrete wavelet transform," *Int. J. Eng. Trends Technol.*, vol. 63, no. 1, pp. 32–39, 2018.
28. S. Kadambe, R. Murray, and G. Paye Boudreaux-Bartels, "Wavelet transform-based QRS complex detector," *IEEE Trans. Biomed. Eng.*, vol. 46, no. 7, pp. 838–848, 1999.
29. S. Z. Mahmoodabadi, A. Ahmadian, and M. D. Abolhasani, "ECG feature extraction using daubechies wavelets," *Proc. 5th IASTED Int. Conf. Vis. Imaging, Image Process. VIIP 2005, no. January 2005*, pp. 343–348, 2005.
30. V. K. Sudarshan *et al.*, "Automated diagnosis of congestive heart failure using dual tree complex wavelet transform and statistical features extracted from 2 s of ECG signals," *Comput. Biol. Med.*, vol. 83, no. January, pp. 48–58, 2017.
31. L. Lu, J. Yan, and C. W. de Silva, "Feature selection for ECG signal processing using improved genetic algorithm and empirical mode decomposition," *Meas. J. Int. Meas. Confed.*, vol. 94, pp. 372–381, 2016.
32. Z. X. Zhang, S. H. Lee, and J. S. Lim, "Comparison of feature selection methods in ECG signal classification," *Proc. 4th Int. Conf. Ubiquitous Inf. Manag. Commun. ICUIMC 10*, pp. 502–506, 2010.

33. M. Ayar and S. Sabamoniri, "An ECG-based feature selection and heartbeat classification model using a hybrid heuristic algorithm," *Informatics Med. Unlocked*, vol. 13, no. March, pp. 167–175, 2018.

34. S. K. Kim, C. Y. Yeun, E. Damiani, and N. W. Lo, "A machine learning framework for biometric authentication using electrocardiogram," *IEEE Access*, vol. 7, pp. 94858–94868, 2019.

35. E. Al Alkeem *et al.*, "An enhanced electrocardiogram biometric authentication system using machine learning," *IEEE Access*, vol. 7, pp. 123069–123075, 2019.

36. L. Biel, O. Pettersson, L. Philipson, and P. Wide, "ECG analysis: A new approach in human identification," *IEEE Trans. Instrum. Meas.*, vol. 50, no. 3, pp. 808–812, 2001.

37. H. J. Kim and J. S. Lim, "Study on a biometric authentication model based on ECG using a fuzzy neural network," *IOP Conf. Ser. Mater. Sci. Eng.*, vol. 317, no. 1, pp. 1–10, 2018.

38. C. E. Shannon, "A mathematical theory of communication," *Bell Syst. Tech. J.*, vol. 27, no. 4, pp. 623–656, 1948.

39. A. M. Ermolaev, "Sparse autoencoder," *J. Phys. B At. Mol. Opt. Phys.*, vol. 31, no. 3, pp. 1–19, 1998.

40. E. Ihsanto, K. Ramli, D. Sudiana, and T. S. Gunawan, "Fast and accurate algorithm for ECG authentication using residual depthwise separable convolutional neural networks," *Appl. Sci.*, vol. 10, no. 9, pp. 1–15, 2020.

41. P. S. Hamilton and W. J. Tompkins, "Quantitative Investigation of QRS Detection Rules Using the MIT/BIH Arrhythmia Database," *IEEE Trans. Biomed. Eng.*, vol. BME-33, no. 12, pp. 1157–1165, 2007.

42. D. Belo, N. Bento, H. Silva, A. Fred, and H. Gamboa, "ECG biometrics using deep learning and relative score threshold classification," *Sensors (Switzerland)*, vol. 20, no. 15, pp. 1–20, 2020.

6 False Media Detection by Using Deep Learning

Renu Dalal, Manju Khari, Archit Garg,
Dheeraj Gupta, and Anubhav Gautam
Ambedkar Institute of Advanced Communication
Technologies & Research

CONTENTS

6.1 INTRODUCTION

In the modern age of media and technology, millions of people share data like images, videos or any other multimedia through the internet, which may or may not be true as it looks or sounds. And most of the time these multimedia do not have genuine content; they are modified to create a false impression in the mind of the audience for personal benefits. The circulation of false media, which can also be termed as deep fakes, is done to gain political or social advantages most of the time, which may severely influence an individual, society and community so the nation in a disastrous manner. These can cause political conflicts between different parties. Face re-enactment permits the author to alter the goal phrases, produce image-realistic photographs and videos. Technology can potentially be applied to many purposes; there is a very significant social implication in the malicious use of automated re-enactment. Therefore, to differentiate original images and videos from the changed ones, it is necessary to establish detection techniques.

Deep fakes are dummy videos or photos that sound and look just like the original thing. Deep fakes means manipulated videos or other digital images generated by advanced artificial intelligence (AI) that generate simulated photos and sounds that seem to be genuine. Deep fakes refer to manipulated videos or other digital images

DOI: 10.1201/9781003138068-6

created by advanced AI that produce real images and voices that appear to be simulated. Deep learning is "a branch of AI" which refers to algorithms that can learn which make good decisions on their own. But the downside of that is "the system will be used to make us think that something is genuine if it is not."

Even after a lot of work in this field, the community is still struggling to deal with deep fakes in an effective way. A harsh truth is that there are very few applications to detect originality of a multimedia video or images, whereas lots of applications are available to make fake ones. So, there is a very strong need for a system to detect fake media, especially videos. Today, in their spare time, anybody can download deeply fake apps and the same realistic fake image. Deep fakes have so far been limited to inexperienced hobbyists putting the actors' faces on the bodies of multiple people and making the politician look odd. One machine learning (ML) model trains a data set and then generates video forgeries while the other tries to identify forgeries. The forger produces forgeries before the other ML model can spot forgeries. The larger the collection of training results, the simpler it is for the forger to construct a convincing deep error. That is why recordings of ex-presidents and Hollywood celebrities have often been included in this early, first wave of in-depth fakes—there is a lot of readily available video clips to train the forger.

Camera applications have gotten more and more complex. Users can elongate knees, erase pimples, add ears, and now some can even make fake videos that look really trustworthy. The technology used to generate such interactive content has rapidly become available to the public, and is named "Deep fakes." How dangerous deep fake can be? Instead of helping everyone, AI-based technology has drawbacks that impact different communities in our culture. Apart from producing false news and manipulation, deep-fake is often used for porn retribution to defame prominent celebrities and as soon as fake videos go viral people first assume, and keep forwarding with others makes the targeted person ashamed to witness such strange activities and also it was seen so many cases related to this in which many people have committed suicide.

Face re-enactment: Face re-enactment evolving conditional face reconstruction task that seeks to accomplish two tasks simultaneously: (i) move the source face form (e.g., facial expression and pose) to the target face; and (ii) maintain the appearance and identification of the target face. In recent years, it has drawn tremendous academic commitment over the years due to its realistic principles in virtual reality and culture. Substitute these methods the face or facial output of a current target image or video with a different face or output from a source picture or video, and a new outcome that looks natural. People use the re-enacted footage to delineate a person saying things he has not said in his life. Such a matter of videos spreading across the Internet to a billion uninformed viewers will lead to a loss of information, chaos and uncertainty on a broad scale. With only little work completed in the past detection, there is an immediate need to establish methods that can be used to diagnose such changes.

The aim of the re-enactment of the face is to animate the source picture to another pose-and-expression given by the driving picture. For example, Face2Face animates the facial expression of the source video using the rendered image. The job of shifting full 3D head location, rotation of head, facial expression, eye focus and eye blinking from the moving actor to the source actor image. Both of these approaches, though, include a vast number of photographs of a specific identity for preparation and only re-enact the face of a specific identity. In this process, on the other hand, this does not have this restriction and is capable of re-enacting any identity provided to a single picture without the need for retraining or fine-tuning.

6.2 RELATED WORK

Image retrieval [1] is generally a function of time cluster of goal frames and a new matching picture method that incorporates presence and motion to pick candidate frames from the source video, while a face transition uses a 2D warping technique that maintains the user's identity. In the proposed framework, excels in simplicity as it does not depend on a three-dimensional face model is stable. Under head motion which does not involve comparable source which targets output. Two approaches used for making hyper-realistic fabricated videos: (a) Deep-fake (b) Face2Face.

Visual material has become the main thing of information, as seen in the billions of photographs and videos that are exchanged and posted every day. This has contributed to a rise in improvements in photographs and videos to generate them more spatially and noticeable for audiences all over the world. Some of these modifications are basic, like copy-moving, and can be quickly detected, whereas other complex modifications, such as re-enactment, are difficult to detect. Re-enactment improvements allow the source to modify the target facial expressions and to produce image-realistic photographs and videos where such improvements are difficult to identify [2]. Important work has been undertaken to produce those photographs and videos. However, the identification of such improvements also needs study.

The growing complexity of mobile technologies and the ever-increasing impact of platform or photo sharing platforms have made the production and dissemination of videos easier than just before. Until lately, the number of false videos and its degree of realistic constraint by the shortage of especially technical checking equipment, the high demand for domain knowledge [3], and it involved a complicated and time-consuming process. In upcoming months, a machine learning free software platform has made it possible to create convincing face swaps in videos that left little signs of deception in what is known as "deep fake" images [4].

New developments in deep generative networks have dramatically enhanced precision and reliability in the development of accurate dupe-face pictures. In this chapter, new approach for releasing fake face videos produced by neural networks. The algorithm is largely based on the detection of eye movement in images, a physiological characteristic which is not suitable for being represented in the synthesized dupe images. The device is validated using a blinking of a detection dataset and also demonstrates promising performance when detecting [5] Deep-Fake-generated images. It is easier to produce, accurate and produce high-resolution images. However, attackers may use these techniques to produce true yet fake images to injure others, bypass image detection algorithms or bypass the recognition of image classifiers. Here, variety of approaches to recognize GANs-created fake images and use pre-processing techniques to improve human-created false image recognition. Proposed methods are based on image content for classification and do not use a meta-data image [6]. Since the Generative Adversarial Network (GAN) may be used to build a rational image, the incorrect usage of these technologies poses hidden concerns. For example, GAN may be used to create a tampered video with actual persons and inappropriate behaviors, to create photographs that are detrimental to a single individual and can also have an effect on personal security. In this chapter, create a Deep Forgery Discriminator (DeepFD) [7] that is used to detect computer-generated images efficiently and effectively. Direct learning of a classifier of binary is surprisingly straightforward since it is tough to identify similar racist features to judge false photographs from various GANs. To solve this problem, follow a contrasting loss in search of the main features of the synthesized images provided by different GANs, and then concatenate the classifier to detect such computer-generated photos [8].

With advances in technology, the probability to create smooth images of human faces for false media is good, taking advantage of the large-scale availability of videos. These composite faces can be used to attack main subjects with characters. The existence of open-source tools and a range of commercial applications offer an incentive to produce false videos of a single target subject in a variety of ways. In this chapter, author tests the generalizability of fake face detection methods through a series of studies to assess the precision of identification. Extensive tests are done using text-based, handcrafted detection techniques and deep learning-based techniques methods to assess the detection methods.

With the emergence of AI and deep learning, false digital content has spread in fresh years. False footage, photos, human speech and videos may be a disturbing and harmful activity that has the potential to modify information and damage faith by providing inaccurate evidence [9–11]. Evidence of authenticity (EoA) of digital media is crucial to the eradication of the epidemic of forged content. Present solutions lack the ability to trace the past and roots of new media. In this chapter, include a solution and a general structure using Ethereum Smart Contracts to map and trace the roots and past of digital content to its origins, even though digital content is copied several times [12–15]. The Smart Contract uses the interplanetary file system hashes used to collect or store content means digital and its metadata. Strategy depends on the quality of video.

6.3 PROPOSED DETECTION ALGORITHM FOR DETECTION OF FAKE MEDIA

Deep learning-based architecture in this research to detect re-enacted frames created using the Face2Face re-enactment technique. In combination with a multi stream network, the proposed approach uses RGB frames for enhanced retrieval of localized facial anomalies and patterns of noise that was proposed by the re-enactment. In order to promote balanced training of the suggested multi-stream network, loss feature is also present in this chapter. By using dedicated streams that discover their respective regional artifacts, the network captures local facial artifacts. The dependence between the regions defines the full-face stream. The suggested network can distinguish highly compressed frames with a relatively minute drop in output compared to existing methods through a combination of learning of regional and full-face objects.

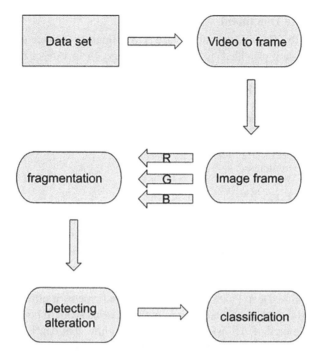

FIGURE 6.1 Flow chart for proposed work.

The flow of the complete process can be understood by the Flow Chart in Figure 6.1. The data set for face re-enactment was created by collecting open-source images and videos of politicians and famous people. Then videos were processed to get frames out of videos, open-source applications, and algorithms were used to create the morphed image of the images collected from the videos, several different apps were used to get variations in the dataset.

Then each image was processed and RGB frames were separated and extracted from the image, for each of these frame analysis was done separately by fragmentation and other steps, then the outcomes from each of these was analyzed to detect the alterations, the result from this step gives a score which is then used to classify the image as a fake image or a real image to detect a fake video the video was first processed to extract frames out of the video and then the frames were processed, by the model, then the mean of the result for all the frames was used to decide if the video is fake or real. This process is called bagging; this improved the efficacy of the output.

6.3.1 BACKGROUND OF PROPOSED WORK

Machine Learning: Machine learning is an implementation of AI that allows systems have the potential to train automatically and build on knowledge without being directly programmed. Machine learning is used in the creation of computer programmes that can gain data and acquire knowledge about themselves. There are following types of ML are: (i) Supervised, (ii) Unsupervised, (iii) Reinforcement.

6.3.1.1   Face-Re Enactment

The purpose of the re-enactment of the face is to animate the source image to another pose-and-expression provided by the driving image. In face re-enactment process, just transfer the face expression of one image into another image so that it will be hard to identify between the morphed image and original image. In face re-enactment uses the algorithm to do this task. GNA is used to perform this task. It has seen so many morphed images on the internet which is difficult to identify between morphed image and original image. And people believe what they see so basically, they see only morphed images and believe it immediately this creates a false impression on the minds of the audience.

6.3.1.2   Deep-Fake

Deep Fake (also read deep fake) is a form of AI used to construct believable images, audio and video cheating. Actually, deep-face is created by two things: one is a generator and the other one is discriminator. Deep Fakes are dummy videos or photos that sound and look just like the original thing. Deep Fakes means manipulated videos or other images generated by advanced AI that generate simulated photos and voice that seem to be genuine. Deep Fakes refer to manipulated videos or other images created by advanced AI that produce real images and voices that seem to be simulated. Deep learning is "a branch of AI" which refers to algorithms that can learn which make good decisions on their own. But the downside of that is "the system will be used to make us think that something is genuine if it is not." Today in spare time everyone could create deep-fake images. So, we have to be very cautious because these fake images could destroy someone's life because it makes people believe that it is not fake, it is a real image or video.

6.3.1.3   Generative Adversarial Network

GANs [10] are an interesting new breakthrough for ML. GANs are generative models: they create new data instances that are close to the training data. GANs are used to produce pictures that look like photos of human beings, even if they do not belong to any actual person.

6.3.2   Experimental Evaluation of Proposed Work

A bunch of images was used for experimentation with the model, one of which was a morphed image of US President Donald Trump (Figure 6.3). It was morphed using the source image (Figure 6.1) to overlap the real image of Mr. Donald Trump (Figure 6.2). The proposed algorithm was used to detect if the image in Figure 6.3 is real or fake. Figure 6.4 depicts the Morphed image.

This is a result which is produced by an algorithm. Actually, it is a deep-fake image. The following figures were the result of extraction of RGB channels (Figure 6.5).

The model was applied on all three frames to detect the alterations, and the output score from all three was evaluated to classify as fake or real, the output was as expected, this is a fake image.

FIGURE 6.2   Source image.

FIGURE 6.3   Target image.

FIGURE 6.4   Morphed image.

FIGURE 6.5   Extracted RGB frames, red, green and blue channels (left to tight).

6.3.2.1 Output

This is a sample output which was observed from the process for creating deep-fake data. Source image which was used for face-re-enactment (Figure 6.6).

Target image which was used for face re-enactment (Figure 6.7). The output after creating the deep-fake of these images can be seen in Figure 6.8. It is really hard to detect with human sight that if the produced image is really a fake one. Result of detection was as expected, and the image in Figure 6.8 was classified as a fake image.

FIGURE 6.6   Source image.

FIGURE 6.7   Target image.

FIGURE 6.8   Produced fake image.

6.4   CONCLUSION AND FUTURE WORK

In this chapter, we introduced a face re-enactment challenge aimed at face expressions from the source human being to the target human being while preserving identification and posing accuracy to the reference images. One-shot face re-enactment model that requires only one image input from the target face is capable of passing the expression of any source face to the target face. This method only takes one picture of any identification, and target facial characteristics, and is capable of creating a HD-quality re-enacted profile image with the original identification but with the target appearance. Identity, facial posture and expression of the image of the face input and of the target mark. The proposed approach using a data set to prove that the proposed model is capable of producing face images with a higher visual quality than the approaches used.

In future work, keep improving proposed model to fill this gap, this gap means we will consider eye movement in face re-enactment so that we will have the finest deep fake images because of this it will be very hard to find errors in an image. Proposed re-enactment images are still not good so to improve these images and will choose an appropriate training approach. It means, first select less resolution image and then image by image and will keep adding the rest to get high quality or also enhance high-definition quality of image. So, by doing this, can be maintained the training process so that it will enhance high quality image with better pixel resolution and it is expecting that proposed toil will enable users to accomplish more productive and productive face re-enactment practice. And new method can be quickly translated to other realms, such as motion migration or body position migration.

REFERENCES

1. Garrido, Pablo, *et al.* "Automatic face reenactment." In *Proceedings of the IEEE Conference on Computer Vision and Pattern Recognition.* 2014.
2. Afchar, Darius, *et al.* "MesoNet: a compact facial video forgery detection network." In *2018 IEEE International Workshop on Information Forensics and Security (WIFS).* IEEE, 2018.
3. Jiang, Liming, *et al.* "Deeperforensics-1.0: a large-scale dataset for real-world face forgery detection." In *2020 IEEE/CVF Conference on Computer Vision and Pattern Recognition (CVPR).* IEEE, 2020.
4. Güera, David, and Edward J. Delp. "Deepfake video detection using recurrent neural networks." In *2018 15th IEEE International Conference on Advanced Video and Signal Based Surveillance (AVSS).* IEEE, 2018.
5. Li, Yuezun, Ming-Ching Chang, and Siwei Lyu. "In ictu oculi: Exposing AI generated fake face videos by detecting eye blinking." *arXiv* preprint arXiv:1806.02877 (2018).
6. Tariq, Shahroz, *et al.* "Detecting both machine and human created fake face images in the wild." In *Proceedings of the 2nd International Workshop on Multimedia Privacy and Security.* 2018.
7. Hsu, Chih-Chung, Chia-Yen Lee, and Yi-Xiu Zhuang. "Learning to detect fake face images in the wild." In *2018 International Symposium on Computer, Consumer and Control (IS3C).* IEEE, 2018.
8. Hasan, Haya R., and Khaled Salah. "Combating deepfake videos using blockchain and smart contracts." *IEEE Access* 7(2019):41596–41606.
9. Agarwal, S., H. Farid, Y. Gu, M. He, K. Nagano, and H. Li. "Protecting world leaders against deep fakes." In *Proceedings of the IEEE Conference on Computer Vision and Pattern Recognition Workshops,* pages 38–45, 2019.
10. Agarwal, A., R. Singh, M. Vatsa, and N. Ratha". "Are image agnostic universal adversarial perturbations for face recognition difficult to detect?" In *IEEE 9th International Conference on Biometrics Theory, Applications and Systems,* pages 1–7, 2018.
11. Afchar, D., V. Nozick, J. Yamagishi, and I. Echizen. "Mesonet: a compact facial video forgery detection network." In *IEEE International Workshop on Information Forensics and Security,* pages 1–7, 2018.

12. Goswami, G., A. Agarwal, N. Ratha, R. Singh, and M. Vatsa. "Detecting and mitigating adversarial perturbations for robust face recognition." *International Journal of Computer Vision*, 127(6):719–742, 2019.

13. Bharati, A, R. Singh, M. Vatsa, and K. W. Bowyer. "Detecting facial retouching using supervised deep learning." *IEEE Transactions on Information Forensics and Security*, 11(9):1903–1913, 2016.

14. Khari, M., R. Dalal, U. Misra, and A. Kumar. "AndroSet: An Automated Tool to Create Datasets for Android Malware Detection and Functioning with WoT." *Smart Innovation of Web of Things*, 187, 2020.

15. Khari, M., R. Dalal, and P. Rohilla. "Extended Paradigms for Botnets with WoT Applications: A Review." *Smart Innovation of Web of Things*, 105, 2020.

7 Evaluation of Text-Summarization Technique

Manju Khari, Renu Dalal, Arush Sharma, and Avinash Dubey
Ambedkar Institute of Advanced Communication Technologies & Research

CONTENTS

7.1 INTRODUCTION

In the modern world, data are the new fuel, not oil. The success of data-driven tech companies has proven that turning data into profit is monumental. However, managing data is not as easy as it sounds. Data do not live in spreadsheets, and infrastructure is required to store and retrieve it. The rapid growth of the internet and its massive influence on every sector of the economy have led to the generation of an enormous volume of information. It is impractical to assume that a human would be

DOI: 10.1201/9781003138068-7

FIGURE 7.1   Process of automatic text summarization.

able to go through hundreds of pages of documents, even if essential. Automatic text summarization is one widely accepted solution. Automatic text summarization is a data science problem of creating accurate, concise, and fluent summaries from large documents. It is meant to assist us in consuming relevant intent faster by providing critical information. Depending on the type of system and record document, its categories are Extractive and Abstractive.

The document required to be summarized is taken as input, pre-processing performs operations like tokenization, removal of stop words, etc., after that every sentence is scored between one and zero based on its importance. Sentences with the highest rank form the summary. Abstractive summarization is when the main points are rewritten using new words, similar to when a human is provided text to summarize. Abstractive summaries try to create coherent statements and eliminate redundancies by creating entirely new sentences. Procedure for automatic text summarization is shown in Figure 7.1. This technique, however, is still not at par with human translation. It is also challenging to develop, making extractive yet a popular option. Extractive summaries are taken directly from the original document and presented in a readable way. It does not contain any rephrasing or coherence. It merely throws out essential parts of the input text. Various representations and scoring techniques are used to identify crucial sentences directly in the source [1, 2]. This makes the scoring process use the core of how efficient the solution is.

This chapter compares five summarising techniques, which are popular and widely used in single document shortening. They are termed as frequency, Term Frequency-Inverse Document Frequency (TF-IDF), Text rank, Summa and Sentence Embeddings. Term frequency based on the bag of words model of representation simply uses the keyword's frequency as its score. The occurrence of key-phrases decides their selection in summary. TF-IDF tries to improve upon the shortcomings of the frequency method. It provides a logarithmic factor that reduces the weight of frequently occurring terms. This makes the selection of common stop words less likely. TextRank is one of the most popular techniques that employ the page rank algorithms probability scoring method. It switches the nodes from web pages to documents. Summa [3] is a variation of Textank that changes the computation of weights and optimizes for better results. Sentence Embeddings uses vector representation to compare semantic relations between words. A quantitative evaluation is performed on the dataset of news articles. The precision and recall provided by ROUGE [4–6] are measured. This evaluation is done by counting the number of sentences selected by the system and are also present in the human goal standard summary. The performance relative to the genre of the article is also noted. It should also be considered that the evaluation of these methods and their accounts on varying genres has not been done.

The structure of the paper is as follows. Section 7.2 introduces algorithms and how each of them measures words and sentences. Section 7.3 analyses the data set used, source and pre-processing. Finally, Section 7.4 presents the future scope and results of the quantitative evaluation.

7.2 RELATED WORK

Extraction-based methods in text Summarization require a technique to score the unit (words or phrases), which has to be extracted from the original document. The Unit can be words or sentences occurring in the paper. Word scoring refers to assigning quantitative value to its importance and pulling out only the keywords. Sentence scoring, on the other hand, fetches the entire sentence.

7.2.1 TERM FREQUENCY (WORD FREQUENCY)

The word frequency method scores the words based on their occurrences. Frequently occurring words are scored higher and are more likely to be in summary. These word scores can then be added to prescribe a sentence score. To eliminate long sentences from being always selected, the score is divided by the frequency of terms in the sentence. These sentences are selected and arranged in their chronological order to get the summary. The word score refers to the number of times a given word occurs in the provided document. This method of representation is called the bag of words [7–9]. The sentence score is an extension of the word score where the word scores of the words present in the sentence are added and the mean is calculated.

$$\text{Word score} = \text{world frequency} \tag{7.1}$$

$$\text{Sentence score} = \text{world frequency/of world} \tag{7.2}$$

7.2.2 TERM FREQUENCY-INVERSE DOCUMENT FREQUENCY

Term frequency incorrectly emphasizes on commonly occurring words which may not contribute to the overall meaning. Hence, inverse document frequency provides a factor that reduces the weight of the pieces that occur frequently and increases the value of times, which happens rarely. Here, it is assumed that rarely occurring words are relatively more important. The Inverse Document Frequency (IDF) is a logarithmically scaled fraction to measure the amount of knowledge provided by the word. The TF-IDF is a product of the term frequency and the IDF to define the importance of the keyword or the phrase within the original document.

$$tfidf(t,d,D) = tf(t,d).idf(t,D) \tag{7.3}$$

$$tf(t,d) = 0.5 + 0.5 \frac{f_{t,d}}{\max\{f_{t'} : t' \in d\}} \tag{7.4}$$

$$idf(t,D) = \log \frac{N}{|\{d \in D : t \in d\}|} \tag{7.5}$$

N: total number of pieces in the data set.

The sentence score is an extension of the word score where the word scores of the words present in the sentence are added and the mean is calculated. IDF provides us information about the relevance of a particular word in the documents if it is rare or common in the document. IDF gives a better understanding of the reference document as the generated summary is much more thorough and closer to the actual document. Term frequency means the number of times a word occurs in a specific document. To eliminate bias in larger documents augmented frequency is used, which means raw frequency divided by the raw frequency of that word which occurs the most in a given document. $|\{d \in D : t \in d\}|$: number of pieces where t is the term appears. If the term is not in the corpus, this will lead to a division-by-zero. It is, therefore, common to adjust the denominator to $1+|\{d \in D: t \in d\}|$.

7.2.3 TEXT RANK

A Text Rank method is derived from the popular page rank algorithm, which is mainly used for ranking pages on the web. The PageRank scores linked web pages on the probability of a user visiting that page called the PageRrank score. The Text Rank algorithm splits the text into sentences and represents them as vectors. The similarities of the vectors are calculated. They are then saved in a matrix. This matrix is then made into a graph with sentences and scores as vertices and edges, respectively. Depending on the requirement for the size of the summary, a threshold score is set, and all penalties above it are selected.

The first step would be to link all the text contained in the articles in sequence. Then divide the text into individual sentences. Further, find vector representation for every single sentence. Figure 7.2 describes text rank implementation. Similarities between sentence vectors calculated and stored in a matrix. The similarity matrix is then converted into a graph, similarity scores as edges and with sentences as vertices, for sentence rank calculation. At last, a specific number of high-ranked sentences form the final summary.

7.2.4 SUMMA

Summa can be described as a different implementation of the main Text Rank algorithm. This variation is done by changing the way to compute the weight of the edges of the graph in PageRank and the distances between sentences. Variations:

(i) Cosine Distance: It is a metric that is used to contrast texts represented as vectors. The TF-IDF model represents documents as vectors and calculates their cosine between them as a measure of similarity.

(ii) Longest Common Substring: From two sentences, identify the longest common substring and report the similarity to be its length.

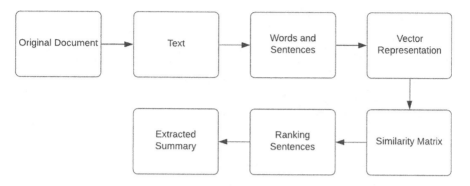

FIGURE 7.2 Implementation of text rank.

7.2.5 SENTENCE EMBEDDINGS

Word embedding represents words in an N-dimensional vector space such that semantically similar or related terms are closer to each other. To get the sentence representation, all the word vectors are added to form a word centroid, and the similarity is measured through the centroid distance. This method can be extended to paragraphs and documents. However, it completely disregards the sequence. All the sentences are clustered using the K-Means Algorithm; each cluster of sentence Embedding can be interpreted as similar sentences, which can be represented by one sentence in summary. This sentence is referred to as a candidate sentence. Candidate sentences corresponding to each cluster are ordered to form the outline. Its original position in the document determines the order.

7.3 EMPIRICAL EVIDENCE

7.3.1 CORPUS

The BBC summary data set is a collection of 1714 news articles (510 Business, 386 Entertainment, 417 Politics, 401 Technology) of BBC from 2004 to 2005. This data set was created using a data set used for data categorization that contains 2,225 documents from the BBC News website containing stories from topical areas used in [4] whose all rights, including copyright, in the original articles' content are owned by the BBC (http://mlg.ucd.ie/datasets/bbc.html). It is obtained from Kaggle, available under the name of BBC news summary, published in 2018.

7.3.2 PRE-PROCESSING

The data set, in its original form, is a directory with two subdirectories. News Articles: containing all the articles divided into genre and Summaries: containing summaries in similar distribution. The individual pieces are stored in separate numbered text files. These files are converted into CSVs (Comma-Separated Values) to be used during implementation. A dictionary is created, and the text in each file is read and stored with the filename as their key. This dictionary is then converted to a panda's data frame and saved in CSV format. The summaries after a similar process are merged with their articles using their associated file numbers. The end lines are replaced with spaces, and data are decoded to UTF8 when required.

Finally, stop words from the text are removed and stemming is performed. Stop words are words deemed irrelevant for the processing of natural language data. They are assumed to not contribute to the overall meaning of the document such as 'a,' 'an,' 'the' etc., and stemming is the process of reducing derived words into their base or root words like consulting, consultant to consult.

7.4   EVALUATION METHOD

ROUGE (Recall-Oriented Understudy for Gisting Evaluation) is used to measure and compare summaries generated by various methods. It is a set of matrices for evaluating the automatic summarization of texts. It reaches human-produced resumes to the one made by the machine, referred to as reference summary and system summary, respectively. Recall, Precision and F-measure are used in ROUGE, where they are calculated as follows:

$$\text{Recall:} \frac{number\_of\_overlapping\_words}{total\_words\_in\_reference\_summary} \qquad (7.6)$$

Recall is the part of the relevant pieces that are successfully retrieved. Recall is also known as true positive rate or sensitivity. Recall is not the only criteria required as it does not take number of irrelevant documents into account.

$$\text{Precision:} \frac{number\_of\_overlapping\_words}{total\_words\_in\_reference\_summary} \qquad (7.7)$$

Precision is the parts of retrieved pieces that are relevant to the query. Precision is also known as positive predictive value. It is the number of correct results returned, divided by all results returned.

$$\text{F-measure:} F_1 = \frac{2}{recall^{-1} + precision^{-1}} = 2. \frac{precision.recall}{precision + recall} \qquad (7.8)$$

F-measure is the harmonic mean of precision and recall. Precision and recall cannot be stated individually because precision and recall can be improved at the expense of other. So, using F-measure gives an overall impression of the performance of the model. ROUGE has different available matrices. This chapter uses three intuitive ones for quantitative measurement. ROUGE 1 which accounts for the overlapping unigrams in the machine summary and the reference summary. ROUGE 2 looks at bigrams, i.e., a pair of words that overlap. ROUGE L measures the longest overlapping subsequence of dishes using the longest common subsequence (LCS). Hence, a predefined value of n-gram is not needed [10, 11].

7.5 EVALUATION

This section presents the results of the quantitative evaluations, details of implementation. The assessment first performed using each article separately, and the abbreviations used are provided in the table for clear understanding.

7.5.1 EVALUATION

Table 7.1 has abbreviations of the metrics, and Table 7.2 provides acronyms for all algorithms. Representations for the metrics that are used in the sections to follow to contrast the various algorithms implemented; representations for the algorithm that have been implemented, numerical representation is used for convenience.

7.5.1.1 TF-IDF

This algorithm is performed in the following steps: (i) all stop words are removed, (ii) stemming is performed, (iii) part of speech tagging is performed in order to obtain nouns, (iv) term frequency and IDF matrix are created, (v) sentence score is given, and the average is calculated, (vi) a threshold score (1.1 * average sentence score) is set, and all sentences above it are extracted, (vii) sentences are arranged in the chronological order of their original text.

7.5.2 METRICS IMPLEMENTATION

The Precision, Recall, and F measure for all articles are added and divided by the number of pieces. This average is used as a representation for the measurement of

TABLE 7.1
Abbreviations

Algorithm	Alg
Rouge 1	Rg1
Rouge 2	Rg2
Average recall	Avg_r
Average precision	Avg_p
Average F-measure	Avg_f

TABLE 7.2
Algorithms

Term Frequency	Alg1
TF-IDF	Alg2
Text Rank	Alg3
Summa	Alg4
Sentence embedding	Alg5

TABLE 7.3

Rg1

Alg	Avg_r	Avg_p	Avg_f
Alg1	0.71	0.34	0.57
Alg2	0.71	0.37	0.57
Alg3	0.62	0.40	0.65
Alg4	0.69	0.41	0.58
Alg5	0.64	0.41	0.60

TABLE 7.4

Rg2

Alg	Avg_r	Avg_p	Avg_f
Alg1	0.63	0.27	0.46
Alg2	0.66	0.31	0.45
Alg3	0.54	0.30	0.45
Alg4	0.57	0.33	0.51
Alg5	0.54	0.35	0.48

TABLE 7.5

RgL

Alg	Avg_f	Avg_p	Avg_r
Alg1	0.63	0.70	0.53
Alg2	0.41	0.58	0.31
Alg3	0.51	0.69	0.41
Alg4	0.48	0.54	0.43
Alg5	0.31	0.49	0.38

the algorithm on the entire data set. Table 7.3 contains the rouge 1 results of the algorithms. Rouge 1 uses single words (no pairs) to compare summaries and its results tend to be higher than rouge 2 and rouge l. Table 7.4 contains the rouge 2 results of the algorithms. Rouge 2 uses bigrams (two adjacent words) to compare reference summaries and system summaries. Table 7.5 contains the rouge l results of the algorithms. Rouge l uses LCS which requires in sequence matches eliminating the need of pre-defined n-gram length.

7.5.3 RESULTS

The results after calculating ROUGE for each algorithm are shown in Table 7.3, Table 7.4 and Table 7.5. Table 7.1 presents Rg1 (unigrams), Table 7.2 presents Rg2 (bigrams) and Table 7.3 presents RgL (longest overlapping subsequence).

a. The best Avg_p in Rg1 is obtained by Alg4 and Alg5, which is 0.41.
b. The highest Avg_f in Rg1 is obtained by Alg3, which is 0.65.
c. The highest Avg_r in Rg2 is obtained by Alg2, which is 0.66.
d. The highest Avg_p in Rg2 is obtained by Alg5, which is 0.35.
e. The highest Avg_f in Rg2 is obtained by Alg4, which is 0.51.
f. The highest Avg_r in RgL is obtained by Alg1, which is 0.53.
g. The highest Avg_p in RgL is obtained by Alg1, which is 0.70.
a. The highest Avg_f in RgL is obtained by Alg1, which is 0.63.
h. The best Avg_r in Rg1 is obtained by Alg1 and Alg2, which is 0.71 (Figure 7.3).
i. Overall, Alg2 is the best performer, followed by Alg1 and Alg3.

Figure 7.3 presents the results of Rg1 on all algorithms where the absolute value of Avg_p is lower than Avg_r and Avg_f. Figure 7.4 and Figure 7.5 present Rg2 and RgL, respectively.

The figure plots the values of recall, precision, and f-measure for single gram rouge (ROUGE-1). The *x*-axis represents each of the algorithms and the *y*-axis denotes the values of the metrics. The metrics are average (mean) values of the value of each article on the data set. The legend of the bar graph is provided on the top right.

The figure plots the values of recall, precision and f-measure for bi-gram rouge (ROUGE-2). These values are lower than rouge-1. The *x*-axis represents each of the algorithms and the *y*-axis denotes the values of the metrics. The metrics are averages (mean) values of the value of each article on the data set. The legend of the bar graph is provided on the top tight.

The figure plots the values of recall, precision and f-measure for rouge-1. It is the longest overlapping subsequence in the summaries being compared. The *x*-axis represents each of the algorithms and the *y*-axis denotes the values of the metrics. The metrics are average (mean) values of the value of each article on the data set. The legend of the bar graph is provided on the top tight.

Rg1

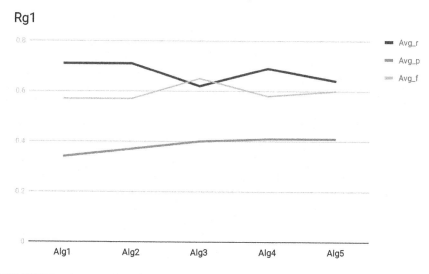

FIGURE 7.3 Avg. metrics of Rg1.

Rg2

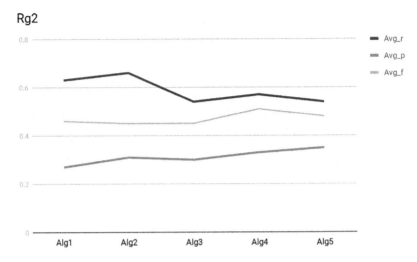

FIGURE 7.4   Avg. metrics of Rg2.

RgL

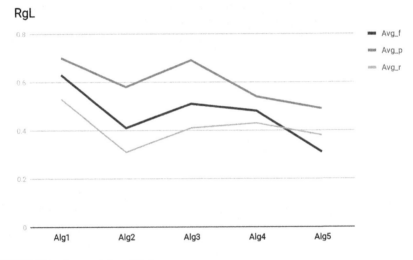

FIGURE 7.5   Avg. metrics of RgL.

7.5.4 Discussion

The following results can be conclusions from the results obtained.

TF-IDF (Alg2) is the best performer on most articles, followed by Word frequency (Alg1) and TextRank (Alg3). The performance of Alg1 and Alg2 shows that the documents depend heavily on the frequency of words. This makes sense because the data set consists of articles from newspapers, i.e., the data are concise and uses well-formed words. News articles are well-written compared to informal writing like tweets or blogs. Hence, TF-IDF, which relies on low-frequency meaningful words, yields the best results.

Both Alg1 and Alg2 are also not computationally intensive, like the other three techniques. This makes them ideal for most situations where documents such as articles or academic writings have to be summarized. They perform well and can be executed on low power machines in a reasonable time.

7.6 CONCLUSION AND FUTURE RESEARCH

This chapter implements and evaluates extractive automatic text summarization techniques. The corpus used consists of articles from four different genres. Each method is implemented in all genres to assess which one performs better according to different matrices. The quantitative comparison is made using ROUGE. Three ROUGE matrices used are Rg1, Rg2, and RgL, which measure overlapping unigrams, bigrams, and sub-sequences. TF-IDF was found to be the overall best performer, followed by Text rank and Word Frequency. Sentence embedding was by far the most computationally intensive method, while word frequency was the least. Rg1 was found to be higher than Rg2 across the data set. In most cases, TF-IDF and Word Frequency can yield the best results with relatively less intensive computation.

It is evident in the results that term frequency and TF-IDF outperform many more complex and nuanced techniques; this is a result of the shortcomings of the approaches. There are several limitations to the research study which can be recommendations for future work.

- There is a lack of large data sets of human summaries to compare them to the machine generated ones. The ROUGE that is used to compare summaries does so by comparing the generated summaries to the ones written by a human. However, there are no large data sets available in such formats.
- This paper contrasts algorithms on News Articles only. Articles tend to be shorter than many long format documents which will affect the precision, recall and F measure scores. The language used in articles is also very specific and condense which will not be the case in many other formats.
- To study how much of an effect pre-processing technique like stemming and removing stop words are making, future work can contrast the two ways by computing the results for both.
- The study uses five of the most popular approaches of text summarization. However, this is not an exhaustive list and other promising techniques like the recurrent neural network are left to be explored.
- Although, ROUGE is the current standard for quantitative comparison of summaries there is also scope for other more efficient ways that may not rely on human summaries to compare two machine generated pieces. Future work involves deeper analysis of new techniques, trying different methods for evaluation, better datasets and curiosity.

REFERENCES

1. Lin, Chin-Yew (2004). Rouge: A package for automatic evaluation of summaries. In Stan Szpakowicz Marie-Francine Moens (Ed.), *Text summarization branches out: Proceedings of the ACL-04 Workshop* (pp. 74–81). Barcelona, Spain: Association for Computational Linguistics.
2. Lloret, Elena, & Palomar, Manuel (2012). Text summarisation in progress: A literature review. *Artificial Intelligence Review*, 37(1), 1–41.
3. Mihalcea, Rada, & Tarau, Paul (2004). TextRank: Bringing order into texts. In *Conference on Empirical Methods in Natural Language Processing*, Barcelona, Spain.
4. Murdock, Vanessa Graham. (2006). Aspects of sentence retrieval. Ph.D. thesis, University of Massachusetts, Amherst.
5. Nenkova, Ani, & McKeown, Kathleen (2011). Automatic summarization. *Foundations and Trends in Information Retrieval*, 5(2–3), 103–233.
6. Nenkova, Ani, & McKeown, Kathleen (2012). A survey of text summarization techniques. In *Mining text data* (pp. 43–76). Springer.
7. Gupta, P., Pendluri, V. S., & Vats. I. (2011). Summarizing text by ranking text units according to shallow linguistic features. In *13th International Conference on Advanced Communication Technology* (pp. 1620–1625).
8. Barrera, Araly, & Verma, Rakesh (2012). Combining syntax and semantics for automatic extractive single-document summarization. In *Proceedings of the 13th International Conference on Computational Linguistics and Intelligent Text Processing* (pp. 366–377). Springer-Verlag.
9. Wei, Yang (2012). Document summarization method based on heterogeneous graph. In *9th International Conference on Fuzzy Systems and Knowledge Discovery (FSKD)* (pp. 1285–1289).
10. Khari, M., Dalal R., Misra U., & Kumar, A. AndroSet: An automated tool to create datasets for android malware detection and functioning with WoT. *Smart Innovation of Web of Things*, 187, 2020.
11. Khari, M., Dalal, R., & Rohilla, P. (2020). Extended paradigms for botnets with WoT applications: A review. *Smart Innovation of Web of Things*, 105.

8 Smart Metro Ticket Management by Using Biometric

Renu Dalal, Manju Khari, Mohammad Nasar Arbab,
Harshit Maheshwari, and Ashirwad Barnwal
Ambedkar Institute of Advanced Communication
Technologies & Research

CONTENTS

8.1   INTRODUCTION

Face Recognition technique is established on a field called Biometric. Biometric access controls are structured approach of certifying a person's integrity on the support of certain physical features, like as fingerprints or facial mien, or other features of a human's behaviour, such as their handwriting pattern or keystroke style. Since biometric structure points a human to basic biological symptoms, it is arduous to replicate them. From the diverse diagnostic approaches, facial recognition is more stable than behavioural methods (key, voice printing). The reason is that physical features often do not change without serious injuries. On the other hand, behavioural patterns may vary because of anxiety, weakness or sickness.

Characteristics of Facial Recognition are as follows: Reproduction: Biological features are inherited, they cannot be varied; therefore it is not possible to reproduce other human's biological features. Availability: Biological factors are a chunk of the human body hence easily accessible and easy to use [1–3].

Applications of Facial Recognition are as follows: Artificial intelligence-based surveillance systems are being used by Law Enforcement agencies which in turn has facial recognition technology as its base. Security associations are using facial recognition to secure their surroundings. Immigration officials use facial recognition for better border control. Internet of Things benefits a lot from facial recognition by permitting extra security measures and programmed control at the residences. A number of smartphones are now using facial recognition to unlock phones which is a resourceful way to protect personal data. Missing children as wells as human trafficking victims are being helped by facial recognition.

Challenges of Facial Recognition

Posture diversity: Head movement: it can be defined by individualistic rotational angles, e.g., angle, turn and curve or camera shift point can lead to major changes in facial expressions and/or shapes and create facial variations in the centre of the subject [4, 5].

Facial changes in expression: Further variations in facial features may be caused by changes in facial features due to different personal circumstances.

Ageing of the face: Changes in appearance can be caused by the ageing of a person's face, and they can affect the entire face-to-face procedure if the time between every photograph is compelling.

Fluctuating lighting conditions: Significant variations in lighting can undermine the performance of facial recognition systems. With low levels of back or front lighting, facial detection and recognition are very difficult to do.

8.1.1 METRO TICKETING SYSTEM

Metro Ticketing Process: The tickets which can be a token or card are bought from the Ticket Counter. The passengers then go towards the metro entry gate that will allow them to enter the metro rail premises. The passenger then holds a token/card near the gate that has an embedded machine [6–9]. The gates and the Tokens/Cards are enabled with Near-Field Communication (NFC) Technology. If the Tokens/Cards are valid the gates will open else it will show an error.

NFC Technology: NFC is a wireless technology. Near-field conversation broadcasts information via electromagnetic radio fields to empower two gadgets to communicate. For it to perform, both gadgets prerequisite contains NFC chips, as execution takes place within small distance [9]. NFC-enabled gadgets must be either physically or very few inches apart from each other in order to transmit data. NFC is derived from the RFID (Radio Frequency Identification) from many communication tools; NFC is one tool which is used to provide imperative contact to digital data and it can be done in three ways.

1.1. Host-card mode: It is a simulation mode, where the device terminal/end point performs like an intelligent contactless chip. The device transmits data to an NFC gadget, e.g., at the ticket verification shop or fee centre. It can work for a variety of objectives: mobile payment, tickets for show or travel, discount coupons, connection control, and etc.

1.2. Reader-mode: In this mode, the device plays an effective work as a contactless chip reader. This mode permits users to receive information or launch operations by touching or tapping user's device next to the NFC tag. For banner, advertisements, souvenirs, bus stop, product packages, NFC identification can be located anyplace on all user social media channels, it increases the scope of uses and availability. Here, the aim is to providing coherent user information to all channels with particular digital data.

1.3. Peer-to-peer mode: Data transfers to both ways that are to and fro between two NFC-connected devices, in this mode. Typically, for sent contact information (vCard) among two smart gadgets, files like photos, videos and make instant cash transfers are transferred.

RFID vs NFC Approach: RFID is the part of Real-Time Localization Systems and it is frequently used for travel purposes in various different areas where market capturing is a key problem: resale, health, transportation and planning, army operations, etc. Therefore, NFC is found in RFID but differs in various ways [10, 11]. Another major difference is that NFC permits for bounded information sharing among the identification and reader. NFC has very short distance range in comparison with RFID (a few inches compared to a few meters). With NFC technique, user can broadcast data among gadgets swiftly and comfortably with a touch operation whether it is for bill payments, for business cards transfer, sharing media or opening our cases and closing machine gates at metro stations.

8.2 RELATED WORK

8.2.1 OBJECT DETECTION

Object detection is known as accurately predicting the idea and places of the objects in each picture. It is capable to give important semantic data for pictures and videos and related to various programs, including image categorization of personality analysis as well as face recognition and self-driving. An explanation for the acquisition problem is to find out where the items are placed in a provided image and what category for each item related to. Traditional object acquisition can be divided mainly into three categories:

Informative region selection: Since a variety of objects can appear in any image position and have a variety of sizes or aspect ratios, this is the original option for scan the entire picture through the multi-scales window. While this perfect approach can capture every location of objects, its disadvantages also exist. Because of high number of voting windows, its cost is high for computers and generates various obsolete windows.

Feature extraction: To see the variety, there is need to bring out the visual elements that can provide strong and solid representation and features are independent. This is because of these traits can trigger the presentation of complex neurons in the human brain. Because of differences in appearance, lighting conditions and its backgrounds, it is complex to design manually a solid feature dictionary to better define all types of objects.

Classification Used: Classifier is necessary to differentiate the identified object in all classes and to make presentations closely related to the sequence, to teach with visual representation.

8.2.2 SIGN LANGUAGE RECOGNITION

Sign language is used by deaf and hard of hearing people communicate information between their community and other people [12]. Computer recognition deals for sign language deals from signal acquisition and continues until text/speech is produced. The vision-based sign language recognition system is able to take pictures of the sign language used with a video camera. An integrated algorithm is included in the face and hand detection system to address the issues of a real-time recognition system. The program removes features of hand movements, face postures and lip movements distinguishly after pre-designed. These characteristics are paired with the signature database and the face database. At the same time, the movement of the lips is extracted by the acquisition of the edge of the image and matched to the oral pattern database. Afterwards transforming the semantics impairment, each recognition outcome was combined for sign-language translation into speech. Most researchers/developers create their own sign-language database. This database can be broken down into digits, alphabets and phrases (simple or complex) [13].

8.2.3 SMART PARKING SYSTEM

In various huge cities and working business places, free parking is extremely complex task to search. As the number of automobiles in today's world grows, it is increasingly stressful and time-consuming to find available parking spaces. A good city strategy for various countries needs smart parking spaces which help drivers to find available/free parking spaces to prevent time delays and avoids problem of traffic. An image processing-based parking program that provides information about unoccupied space to help the driver book the available parking space in their homes to reduce traffic congestion. The system looks at the driver's face and uses it to match the face stored in an existing database during slot bookings. The smart parking system was designed with the help of hardware and software based on the idea of face recognition, and the web application is used by the driver to easily view parking statistics. A major incentive is to renew the parking procedure by decreasing the time needed to park the vehicles [14].

8.3 PROPOSED MODEL FOR METRO TICKETING SYSTEM

The objective of this project is to design a Software Product that enables entry/exit from the metro platform gates using facial recognition, further which also keeps track of the travel history and balance. It Implements the use of facial recognition and cloud database to provide an enhanced and efficiently programmed metro ticketing system. The system demonstrates the installation of a camera at every metro station facility at their respective entry and exit points in the metro rail to help in the computation of the ticket prices [15, 16]. As per the distance (number of stations) travelled by the passenger the respective cost is deducted from their metro account. The whole process is using a programmable database system that makes the implementation easier, faster and uncertainty free.

8.3.1 ARCHITECTURE

8.3.1.1 React JS Application

It is a web-based application that is responsible for interacting with the passenger travelling in the metro it will scan the face while entering and exiting. The metro stations make an HTTP request (sending a message to the server in form a request) after scanning the face and the server then gives an HTTP response (after receiving the request and interpreting is the server then sends/responds with a message) [17, 18]. And according to the response the metro gates open up or show error message. Architecture for the proposed work smart metro ticket management system is shown in Figure 8.1.

8.3.1.2 Image Matching Model

The Image Matching Model gets its data from the database and the server is responsible for analysing the images given by the server and the already existing images in the database and gives an answer to the server according to the analysis done by it. The Image Matching Model uses a Perpetual hashing algorithm that produces a fragment or fingerprint of multiple types of multimedia or in our case the given image. Perceptual hash algorithm functions are analogous if the given features are the same.

8.3.1.3 API Gateway

API Gateway is an API management tool that stays between the client and a set of backend services. The API portal breaks down multiple applications, moves them to the right places, generates feedback and tracks everything. The API port determines all incoming applications and sends them through the API management system, which handles the various tasks required.

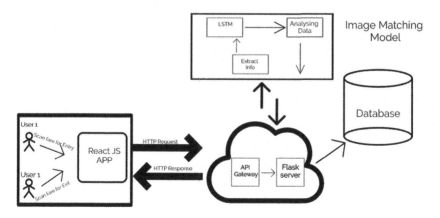

FIGURE 8.1 Architecture for proposed work.

8.3.1.4 Flask Server

It is responsible for handling the login and registration of the passenger and it inter-acts with the database making sure that it gives access to the people who are already registered. It interacts with the Image matching model by taking the analysed result and tells the web application what to do on the particular acquired result.

Database

It consists of the information of the passengers (Name, Email, Balance) and the information of the metro stations. It interacts with the flask server and the image matching model and gives them access to the data that is required by them. The image below shows the database schema that is being used.

8.4 RESULTS AND ANALYSIS

8.4.1 TECHNOLOGY STACK

ReactJS: It is an open-sourced JavaScript library for building user connections or UI components. It is owned and maintained by Facebook and the community of individual developers and organisations. React is used as a basis for creating a single-page and certain mobile applications. React has been used as a basis for developing end-to-end web-based applications as well.

Flask: It is a framework for web application which is written in Python. Flask is built on the basis of Werkzeug WSGI toolkit and template engine of Jinja2. Flask is often called a micro framework. It aims to keep the app's core simple but expandable. Flask does not have a built-in extraction layer for data management and does not have form verification support. Instead, Flask supports extensions to add such purposes to the app.

Android Studio: It is the combined development environment for the Android app development. Based on IntelliJ IDEA, an integrated Java software development platform. For support system development within the Android app, Android Studio utilised a Gradle-based building program, simulator, templates for code and GitHsub integration. Every project in Android Studio has one or more methods with source code and application files [8]. User details, metro station details and travel history for users shown in Figures 8.2–8.4, respectively.

Database:

The Database Schema

Id [PK] serial	created_at timestamp with time zone	updated_at timestamp with time zone	name character varying	email character varying	image text	phone character varying	balance double precision
1	8	2020-04-15 06:49:35.28976+00	2020-04-15 06:49:35.28976+00	Ashirwad Barnwa	ashirwad@gmail.com	https://ima	9560887254	1000
2	9	2020-04-15 06:50:01.132271+00	2020-04-15 06:50:01.132271+00	Nasar Arbab	nasar@gmail.com	https://ima	9521887254	1000
3	10	2020-04-15 06:50:48.293629+00	2020-04-15 06:50:48.293629+00	Priyanka	pinki@gmail.com	https://ima	9013287254	1000
*								

FIGURE 8.2 User details.

	id [PK] serial	created_at timestamp with time zone	updated_at timestamp with time zone	name character varying
1	2	2020-04-15 06:42:30.504501+00	2020-04-15 06:42:51.37951+00	Rajiv Chowk
2	3	2020-04-15 06:43:25.304524+00	2020-04-15 06:43:04.969547+00	Shahdara
3	4	2020-04-15 06:43:16.904391+00	2020-04-15 06:43:09.124484+00	Kashmere Gate
4	5	2020-04-15 06:44:07.864998+00	2020-04-15 06:44:11.444904+00	Inderlok
5	6	2020-04-15 06:44:16.434415+00	2020-04-15 06:44:21.404986+00	Laxmi Nagar
6	7	2020-04-15 06:44:25.944642+00	2020-04-15 06:44:30.764874+00	Shastri Park
*				

FIGURE 8.3 Details for metro stations.

	id [PK] serial	created_at timestamp with time zon	updated_at timestamp with time zone	origin character varying	destination character varying	fare integer	entry_time timestamp without time zone	exit_time timestamp without time zone	user_id integer
1	6	2020-04-15 06:53:23.2	2020-04-15 06:53:25.834448+00	Rajiv Chowk	Kashmere Gate	20	2020-04-15 06:53:30.24065	2020-04-15 06:54:21.894457	10
2	7	2020-04-15 06:53:23.2	2020-04-15 06:53:25.834448+00	Shahdara	Inderlok	40	2020-04-15 06:53:30.24065	2020-04-15 06:54:21.894457	10
3	8	2020-04-15 06:53:23.2	2020-04-15 06:53:25.834448+00	AIIMS	Shahdara	10	2020-04-15 06:53:30.24065	2020-04-15 06:54:21.894457	8
4	9	2020-04-15 06:53:23.2	2020-04-15 06:53:25.834448+00	Rajiv Chowk	INA	60	2020-04-15 06:53:30.24065	2020-04-15 06:54:21.894457	10
5	10	2020-04-15 06:53:23.2	2020-04-15 06:53:25.834448+00	Lokpuri	Dilshad Garden	30	2020-04-15 06:53:30.24065	2020-04-15 06:54:21.894457	9
*									

FIGURE 8.4 Details of travel history.

8.4.2 USE CASE

Working: Database schema use case diagram for proposed work is shown in Figures 8.5 and 8.6, respectively. Figures 8.7 and 8.8 depict the workflow of android application and web application.

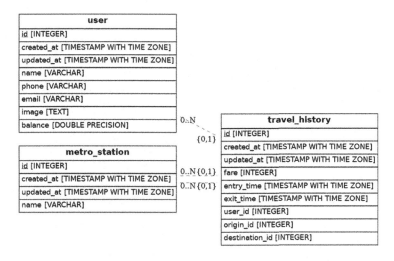

FIGURE 8.5 Database schema for model.

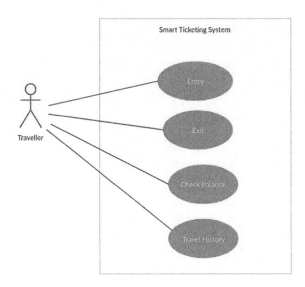

FIGURE 8.6   Use case diagram for model.

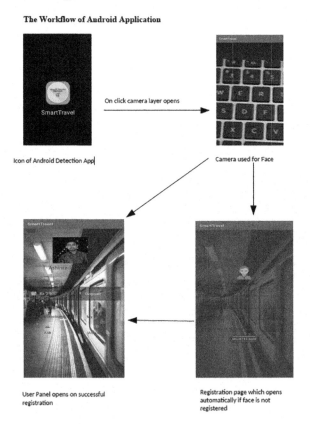

FIGURE 8.7   Work flow for android application.

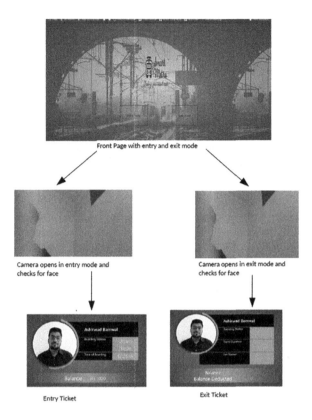

FIGURE 8.8 Work flow for web application.

- The first part of the process includes one-time registration on the user.
- The passenger proceeds towards gates for entry.
- The passenger's face is itself the token/ticket.
- The passenger looks towards the camera and if they have sufficient balance the gates pen.
- While exiting a similar process is followed and the respective fare is deducted from the passenger's balance.
- Apart from this the passenger can recharge, check balance and their past travel history.

8.5 CONCLUSION AND FUTURE SCOPE

Facial recognition is a really interesting subject to study and to develop various projects and able to develop an application with facial recognition as its core. From the result that we have acquired from this project, it can be said that facial recognition can work as an alternate method for ticketing purposes for Metro travel. Since the very beginning of metro, the passengers wanted to travel without any hassle and with a sense of security and with the implementation of this application the passengers now have the luxury of not carrying any tokens or cards and not only that but it also reduces plastic waste. The main aim was to fully automate the ticketing process, so now the passengers and enter, exit, recharge and check their balance just with just scanning their faces.

This work explores only a small part of the science of facial recognition. As a new discipline, there is a great deal and more research and development to do. The following section describes the areas for research that were offshoot or tangential to main objective. Various facial recognition applications are available on the internet but it is not known if they are being used. (i) This technology can also be upgraded for use in other ways such as ATMs, to access private files or other sensitive objects. This could disable some security measures such as passwords and keys. (ii) It takes a lot of time while checking in at the airports but with the implementation of this technology boarding into flights would be much easier as the passenger would no longer need to carry a boarding pass or passport just scan the face and then move to the body scanner. (iii) The conventional method of name calling for attendance in schools can be replaced with facial recognition which in turn would reduce the workload and it would be much easier to maintain the acquired data as well.

REFERENCES

1. https://developer.android.com
2. https://semantic-ui.com
3. https://reactjs.org/
4. https://machinelearningmastery.com/
5. https://docs.aws.amazon.com/
6. https://docs.python.org/3/
7. https://flask-doc.readthedocs.io/en/latest/
8. https://core.ac.uk/download/pdf/26858086.pdf
9. https://www.investopedia.com/terms/n/near-field-communication-nfc.asp
10. https://www.ariamindshare.com
11. https://www.unitag.io/nfc/what-is-nfc
12. https://arxiv.org/pdf/1807.05511.pdf
13. https://www.researchgate.netpublication262187093Sign_language_recognition_Ste_ of_the_art
14. https://www.ijrte.org/wp-content/uploads/papers/v8i1C2A11680581C219.pdf
15. Mednieks, Z. R., Dornin, L., Meike, G. B., & Nakamura, M. (2012). *Programming android*. O'Reilly Media, Inc. Sebastopol, Tokyo.
16. Rosner, T. (2013). *Learning AWS OpsWorks*. Packt Publishing Ltd. Birmingham, Mumbai.
17. Khari, M., Dalal, R., Misra, U., & A. Kumar. AndroSet: An Automated Tool to Create Datasets for Android Malware Detection and Functioning with WoT. *Smart Innovation of Web of Things*, 187–206, 2020.
18. Khari, M., Dalal, R., & Rohilla, P. Extended Paradigms for Botnets with WoT Applications: A Review. *Smart Innovation of Web of Things*, 105–122, 2020.

9 Internet of Things: Security Issues, Challenges and Its Applications

Sachin Dhawan
Ambedkar Institute of Advanced Communication
Technologies and Research

Dr. Rashmi Gupta
NSUT East

Arun Kumar Rana
Panipat Institute of Engineering and Technology

CONTENTS

DOI: 10.1201/9781003138068-9

111

9.1 INTRODUCTION

In the current era, the Internet of Things (IoT) has gained tremendous consideration. Kevin Ashton first proposed the concept of IoT in 1999. The IoT is the most important topic by which everything is interconnected like articles, administrations, people and gadgets that can exchange and interact knowledge and data in various regions and applications to achieve goals. In various fields such as agribusiness, medical treatment, energy production and circulation, and numerous different territories, IoT can be performed, expecting items to interface with the help of web to conduct various tasks like business shrewdly without human inclusion. IP address is used to define the IoT area, but each substance has a special reference to that locale by which it is known. Due to rapid developments in portable correspondence, RFID, WSN and distributed computing, IoT gadget interchanges are becoming more useful than they were previously. The World of IoT incorporates an enormous assortment of gadgets that incorporate PDAs, PCs, PDAs, PCs and other hand-held inserted gadgets. The IoT gadgets depend on practical sensors and remote correspondence frameworks to impart with one another and move important data to the concentrated framework. The data from IoT gadgets are additionally prepared in the incorporated framework and conveyed to the proposed objections. With the fast development of correspondence and web innovation, our everyday schedules are more focused on virtual world [1]. Individuals can visit and can do any other tasks like shopping, any work, though people live in reality. In this way, it is hard to replace all the human exercises with the completely robotized living. The IoT has effectively incorporated the anecdotal space and this presents reality on a similar stage. The significant focuses of IoT are simply the arrangement of a shrewd climate and reluctant autonomous gadgets, for example, brilliant living, savvy things, keen well-being and savvy urban areas among others [2]. These days the selection pace of the IoT gadgets is exceptionally high, an ever increasing number of gadgets are associated through the web. As indicated by evaluation [3,4], due to IoT devices, total income produced is around 700 billion euros with the help of 30 billion connected things with 200 billion associations. Presently in Asia, there are 15 billion gadgets that are relied upon to arrive at 24 billion constantly 2020. In next few years, our life style and plans of action will be dependent on IoT. It will allow individuals and gadgets to impart whenever, wherever [5], with any gadget utilizing any organization and any assistance. The main aim of IoT is to make world superior for individuals in future. The complete working of IoT or definition of IoT is given by Figure 9.1. Shockingly, most of these gadgets and applications are not intended to deal with the protection assaults and it builds a ton of security and security issues in the IoT organizations, for example, classification, confirmation, information respectability, access control, mystery, and so on [6]. On consistently, the IoT gadgets are focused by aggressors and interlopers. An examination reveals that 70% of the IoT gadgets are anything but difficult to assault. In this manner, a proficient system is incredibly expected to make sure about the gadgets associated with the internet against hackers and intruders [7–9].

The IoT is a remarkable overall data network involving Internet-related articles, for instance, Radio repeat recognizing bits of proof, sensors, actuators, similarly as various instruments and sagacious machines that are transforming into a fundamental piece of things to come Internet. All through the latest decade, we have seen innumerable IoT courses of action made by new organizations, little and medium undertakings, huge ventures, insightful investigation foundations (for instance, universities), and private and public assessment affiliations progressing into the market.

9.2 ARCHITECTURE OF IoT

IoT architecture contains three important layers. These three layers are shown in Figure 9.2.

9.2.1 PERCEPTION LAYER

Most important layer of IoT is perception layer or also known as Sensor layer in IoT. Sensor layer is used for collecting information from climate with the help of various sensors like Temperature Sensors, Humidity Sensors, Pressure Sensors, Proximity Sensors and Level Sensors, etc. After collecting the information from climate, it conveyed that information to the organization layer for further processing [5].

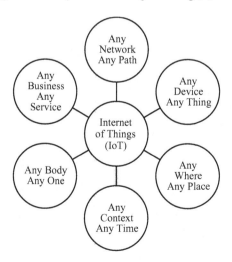

FIGURE 9.1   IoT.

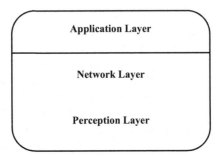

FIGURE 9.2   IoT architecture [2].

9.2.2 NETWORK LAYER

The second main layer of IoT used to pass the information to various other IoT hubs and gadgets with the help of Internet. Wi-Fi, LTE, Bluetooth, 3G and Zigbee, like modern technologies, are used to run internet switching and routing devices. Internet connection makes the path for various IoT gadgets with the help of various different sensors and communicates data from one place to other required places [6].

9.2.3 APPLICATION LAYER

The third layer of IoT that guarantees the uprightness, secrecy, authenticity of the information is known as application layer. In this, the aim of IoT which is the production of savvy conditions is executed [10].

9.3 SECURITY ISSUES AND FEATURES OF IoT

Integrity and availability are mandatory security priorities for the security of IoT confidentiality and should be available for all communications. A sudden growth in the number of IoT devices worldwide would make this IoT security problem volatile in the context of potential permutations. Still, since much of the IoT data is stored in the cloud, this malware does not have useful data to lock.

9.3.1 SECURITY FEATURES OF IoT

It is possible to narrowly divide IoT security into two classes: technical and security complaint [6]. Due to the peculiar and inescapable existence of IoT gadgets, technical problems occur, whereas the protection incitement is identified with the morality and convenience that should be performed to achieve a safe organization. Throughout the creation and present scenario of all IoT gadgets and center points, security should be remembered for IoT [5]. Some security rules are given below.

Secrecy—Ensuring the data is safe and readily available to supported customers is important. This relies on the trading of information and data between various gadgets, which is why it is important to check the accuracy of the information; the information is obtained from the right sender to ensure that the information is not changed due to expected or unforeseen obstruction along the way to contact.

Availability—IoT's vision is to enter any number of keen gadgets as permitted. IoT devices must be visible at any stage whenever they required. Knowledge is not the key modules that are used in the IoT anyway; gadgets and administrations should also be friendly and open to meet IoT standards when needed in a convenient design.

Authentication—Each linked item in the IoT must be so advanced that various articles can be recognized and checked. In addition, things can have to talk spontaneously with various articles now and then (objects that don't have the foggiest idea) [11]. In view of this a strategy is needed for the common confirmation of elements in each IoT correspondence.

Lightweight Solutions—IoT cannot miss any of the security objectives considered before, despite the fact that it could apply exceptional characteristics and specifications to each of them. In any event, in any classification, trustworthiness, accessibility and validation are regarded as a fundamental objective in the protection of each PC or organization.

Heterogeneity—It means the relation of different components with capacity, unpredictability, and merchants that distinguish. The IoT establishes ties between gadget to gadget, human to gadget and human to human, and thus updates the relation between different items and organizations [6]. In IoT, some requirements must be fulfilled that the environment is constantly changing (elements), a system may be linked to a completely new arrangement of gadgets at one time than at any other time. Moreover with appropriate key administration and security conventions, a security ideal cryptography system is needed to ensure security.

Policies—There must be rules and guidelines in the IoT to ensure that all information is secured, organized and shared in a constructive manner, but a mechanism to achieve such arrangements is required to ensure that each aspect updates the principles more critically. In each assist included, Service Level Agreements (SLO) must be clearly recognized. The need for such rules would imply confidence in the IoT model for human customers, which will evolve and adapt in the future.

9.3.2 IoT Security Risks

There are many IoT security risks.

9.3.2.1 Lack of Observance on the Part of IoT Manufacturers

Various IoT gadgets come out practically day by day, all with unfamiliar weaknesses. The essential well-spring of most IoT security issues is that makers do not invest enough energy and assets on security [11].

In this various IoT security risks may involve like weak, guessable, or hardcoded passwords, hardware issues, lack of a protected update component, old and unpatched inserted working frameworks and programming, insecure information move and capacity.

9.3.2.2 Lack of User Knowledge and Awareness

Internet experts have found out how to dodge spam or phishing messages and examine and protect their networks with good passwords on their machines.

In any event, IoT is another breakthrough, and people do not even care much about it. And most of the IoT protection risks are still on the assembly side, so consumers and business cycles can pose greater dangers. The customer's oblivion and lack of commitment to the IoT utility are the main issues of IoT security and challenges. Consequently, everyone is put at risk. More often than not, deceiving a person is the most effortless path to entering an entity. Social design assaults are a sort of IoT protection threat that is mostly overlooked. A programmer focuses on a human being using the IoT, rather than concentrating on devices.

9.3.2.3 Issue of IoT Security in Device Update Management

Shaky programming or firmware is an additional source of IoT security opportunities. While a manufacturer can sell a gadget with the latest programming update, it is virtually impossible that new vulnerabilities will arise. For IoT gadgets, updates are important for maintaining security. Just after fresh vulnerabilities are discovered, they should be refreshed. In any case, without the basic updates, some IoT gadgets continue to be used as contrast and mobile phones or PCs that receive programmed refreshes.

Another possibility is that a device will send its reinforcement out to the cloud during an upgrade and will experience a short holiday. A programmer could take touchy data on the off chance that the association is decoded and the upgrade records are unprotected.

9.3.2.4 Lack of Physical Hardening

Likewise, the lack of real solidification will cause IoT security problems. While some IoT gadgets should have the option of operating self-sufficiently without a customer's intercession, external dangers should really be ensured. These devices can be found in distant areas for a substantial amount of time once in a while and they may be actually altered, such as using a USB streak drive with Malware. The manufacturer begins by guaranteeing the real security of an IoT gadget. In any event, it is a difficult undertaking for manufacturers to develop stable sensors and transmitters in the all-around simple gadgets. Clients are also responsible for genuinely ensuring that IoT gadgets are preserved. If not properly guarded, a savvy motion sensor or a camcorder that sits outside a house could be messed with.

9.3.2.5 Botnet Attacks

A solitary IoT gadget contaminated with malware does not represent any genuine danger; it is an assortment of them that can cut down anything. To play out a botnet assault, a programmer makes a multitude of bots by tainting them with malware and guides them to send a large number of solicitations every second to cut down the objective.

A large part of the turmoil about IoT security started after the Mirai bot assault in 2016. Various Distributed Denial of Service assaults utilizing countless IP cameras, NAS and home switches were contaminated and coordinated to cut down the DNS that offered types of assistance to stages like GitHub, Twitter, Reddit, Netflix and Airbnb.

9.3.2.6 Industrial Espionage and Eavesdropping

In the event that programmers assume control over observation in an area by contaminating IoT gadgets, spying probably will not be the main choice. They can likewise perform such assaults to request emancipate cash.

Accordingly, attacking protection is another noticeable IoT security issue. Spying and interrupting through IoT gadgets is a genuine issue, as many delicate information might be undermined and utilized against its proprietor.

9.3.2.7 Hijacking Your IoT Devices

Ransomware does not pulverize your delicate documents—it blocks admittance to them via encryption. At that point, the programmer who tainted the gadget will request a payoff charge for the unscrambling key opening the records.

9.3.2.8 Data Reliability Risks in Healthcare for IoT Security

With IoT, information is consistently progressing. It is being communicated, put away and prepared. Most IoT gadgets concentrate and gather data from the outer climate.

It tends to be a keen indoor regulator, HVAC, TVs and clinical gadgets. In any case, now and again, these gadgets send the gathered information to the cloud with no encryption.

Accordingly, a programmer can access a clinical IoT gadget, overseeing it and having the option to modify the information it gathers. A controlled clinical IoT gadget can be utilized to impart bogus signs, which thusly can make well-being professionals make moves that may harm the soundness of their patients.

9.3.2.9 Rogue IoT Devices

The fast development in the technology of IoT, IoT gadgets anticipated arriving at 18 billion by 2023. One of the main IoT security dangers and difficulties is having the option to deal with every one of our gadgets and close the border.

Be that as it may, maverick gadgets or fake noxious IoT gadgets are starting to be introduced in made sure about organizations without approval. A maverick gadget

FIGURE 9.3 Raspberry Pi board [5].

replaces a unique one or incorporates as an individual from a gathering to gather or adjust delicate data. These gadgets break the organization border.

Illustration of rouge IoT gadgets can appear as the WiFi Pineapple or Raspberry Pi. These can be transformed into a maverick AP (Access Point), indoor regulator, camcorder, or Man in the Middle and block approaching information correspondences unbeknownst to clients. Different varieties of maverick gadgets may likewise arise later on.

Strangely, the thriller "No problem" was roused by the idea and can fill in as an inquisitive model. In the film, controlling different gadgets in a brilliant home framework, Chucky is a rouge IoT gadget that has become an elevated level danger to individuals' lives.

Figure 9.3 shows the Raspberri Board.

9.3.2.10 Crypto Mining with IoT Bots

Mining digital currency requests huge processing unit assets, and because of this crypto mining with various connected devices another IoT security issue has arisen. This sort of assault includes tainted botnets focused on IoT gadgets, with the objective not to make harm, however, mine digital currency.

The open-source digital money Monero is one of the initial ones to be mined utilizing contaminated IoT gadgets, for example, camcorders. Albeit a camcorder does not have incredible assets to mine digital currency, a multitude of them does.

IoT botnet excavators represent an incredible danger to the crypto market, as they can possibly flood and upset the whole market in a solitary assault.

9.4 IoT CHALLENGES

IoT application data could be new, endeavor, client or entity. This knowledge about the application should be ensured and secret for theft and alteration. The IoT applications, for instance, can store the aftereffects of a well-being or shopping store for patients. The IoT improves the interaction between gadgets, but there are problems with adaptability, usability and reaction time. Safety is where information is shared securely over the internet. The Protection Measure Act could be enforced by government guidelines. Among numerous security issues, the key problems related to the IoT are addressed [12].

(a) Data Privacy: Makers of keen TVs gather information about their clients to dissect their survey propensities so the information gathered by the savvy TVs may have a test for information protection.

(b) Data Security: This is additionally an incredible test. While sending information consistently, it is essential to stow away from noticing gadgets on the web.

(c) Insurance Concerns: For security check, various agencies of insurance introducing IoT gadgets on vehicles gather information about well-being and status of driver to take choices about protection.

(d) Lack of Common Standard: There are numerous norms for IoT gadgets and IoT fabricating ventures. Hence, it is a major test to recognize allowed and non-allowed gadgets associated with the web.

(e) Technological Concerns: Because of the expanded utilization of IoT gadgets, the devices produced by these gadgets are additionally expanding. Thus it is required to build network limit, subsequently, it is additionally a test to store the tremendous measure of information for examination and further last stockpiling.

(f) Security Attacks and System Vulnerabilities: Lots of work have already done in the area of IoT security. The connected work can be isolated into framework security, application security, and organization security [13].

(g) System Security: System security primarily centers around generally speaking IoT framework to recognize diverse security challenges, to plan distinctive security structures and to give legitimate security rules to keep up the security of an organization.

(h) Application security: This works for application of IoT to deal with various issues in security as indicated by situation necessities.

(i) Network security: Network security manages making sure about the IoT correspondence network for correspondence of various IoT gadgets. In the following segment, the security concerns with respect to IoT are talked about. The security assaults are arranged into four expansive classes.

9.5 DIFFERENT TYPES OF ATTACKS AND POSSIBLE SOLUTIONS

The IoT is confronting different sorts of assaults including dynamic assaults and detached assaults that may handily upset the usefulness and cancel the advantages of its administrations. In an inactive assault, an interloper just faculties the hub or may take the data however it never assaults truly. In any case, the dynamic assaults upset the presentation truly. These dynamic assaults are ordered into two further classes that are interior assaults and outside assaults. Such weak assaults can forestall the gadgets to impart adroitly. Thus the security limitations must be applied to keep gadgets from malignant assaults [14]. Various kinds of assault, nature/conduct of assault and danger level of assaults are talked about in this part. Various degrees of assaults are ordered into four kinds as per their conduct and propose potential answers for dangers/assaults. (i) Low-level assault: If an assailant attempts to assault an organization and his assault is not effective. (ii) Medium-level assault: If an assailant/gatecrasher or a busybody is simply tuning into the medium yet do not adjust the respectability of information. (iii) High-level assault: If an assault is carried on an organization and it changes the respectability of information or adjusts the information. (iv) Extremely high-level assault: If a gatecrasher/assailant assaults on an organization by increasing unapproved access and playing out an unlawful activity, making the organization inaccessible, sending mass messages or sticking organization [15–17].

9.6 IoT APPLICATIONS

The principal goals of IoT are simply the arrangement of a savvy climate and hesitant-free gadgets, for example, shrewd living, keen things, brilliant wellbeing and shrewd urban communities among others [2]. The utilizations of IoT in businesses, clinical field, and in home robotization are talked about in the accompanying segment. Figure 9.4 shows the various applications of IoT.

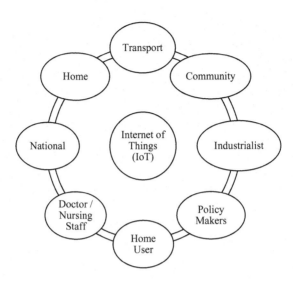

FIGURE 9.4   IoT applications [2].

9.6.1   IoT in Industries

This has given a reasonable occasion to construct huge modern frameworks [18]; in a smart transportation using IoT framework, the approved individual can screen the current area and development of a vehicle. In industrial application smart robotics is the most famous and important application of IoT. In this era, online retail is also growing with the help of human machine interface and IoT. Manufacturing efficiency also increasing and industrial revolution started with the help of IoT. The IoT licenses numerous things to be associated, followed and observed so important data and private information gathered consequently. In IoT climate, the security insurance is a more basic issue when contrasted with conventional organizations since quantities of assaults on IoT are high.

9.6.2   IoT in Personal Medical Devices

The use of the IoT in applications of healthcare has increased across numerous unique cases. Most healthcare IoT programs have concentrated on improving treatment as such with remote surveillance as primary implementations in the telemedicine scope. Tracking, monitoring and management of infrastructure, using RFID and IoT, is a second field where many initiatives occur. This is achieved at the level of equipment used in medical and healthcare services, the level of individuals and the level of non-medical properties. These deployments and use cases, however, are only the beginning and are far from omnipresent at the same time. The IoT gadgets are additionally generally utilized in medical care frameworks for observing and evaluation of disease [19]. IoT devices related to healthcare help to provide medicine to the patient's body. Many electronics devices have been invented which are used as personal medical devices. The market estimation of these gadgets is extended to associate with 30 billion dollars by 2021. On account of medical care, the essential objective is to guarantee the security of organization to keep the protection of patient from malignant assaults. At the point when assailants assault cell phones, they have their predefined objectives [20]. As a rule, their point is to take the data, assault on gadgets to use their assets, or may close down certain applications that are observing patient's condition. There are numerous sorts of assaults on clinical gadgets that incorporate snooping in which security of the patient is spilled, respectability blunder in which the message is being changed, and accessibility issues which incorporate battery depleting assaults. Some network protection dangers identified with security, and wellbeing of clinical information of patient are talked about as follows:

1. PMDs are basic to any errand that utilizes battery power. Henceforth these gadgets must help a restricted encryption. On the off chance that the gadget is a piece of various organizations; at that point secrecy, accessibility, security, and honesty will be at high danger.
2. As PMDs have no validation system for remote correspondence. So the data put away in the gadget might be effectively gotten to by unapproved people.
3. Absence of secure confirmation additionally reveals the gadgets to numerous other security dangers that may prompt vindictive assaults. An unfriendly may dispatch Denial of Service assaults.
4. The information of patient is sent over transmission medium which might be adjusted by unapproved parties; subsequently protection of a patient may misfortune.

9.6.3   SMART HOME USING IoT

The IoT shrewd home administrations are expanding step by step [21]; computerized gadgets can adequately speak with one another using internet system. All shrewd home gadgets are associated with the web in a keen home climate. As the quantity of gadgets increments in the keen home climate, the odds of pernicious assaults likewise increase. On the off chance that brilliant home gadgets are worked autonomously the odds of pernicious assaults additionally diminish. As of now savvy home gadgets can be gotten through the web wherever and whenever. Thus, it builds the odds of noxious assaults on these gadgets.

9.6.4   IoT IN BIOMETRICS

Biometric recognition framework, security and encryption norms are properly joined into different fields for application on a more prominent scale [22,23]. Biometric security system becomes more secure with the innovation of IoT and encourages better confirmation of arrangement for most extreme security guidelines [24].

A. Banking and E-Payment: Internet and various versatile modes like block chain systems and e-trading offices are used.
B. Corporate and Enterprise levels [25]: Biometric recognition system allowed employee access (immediate or distant).

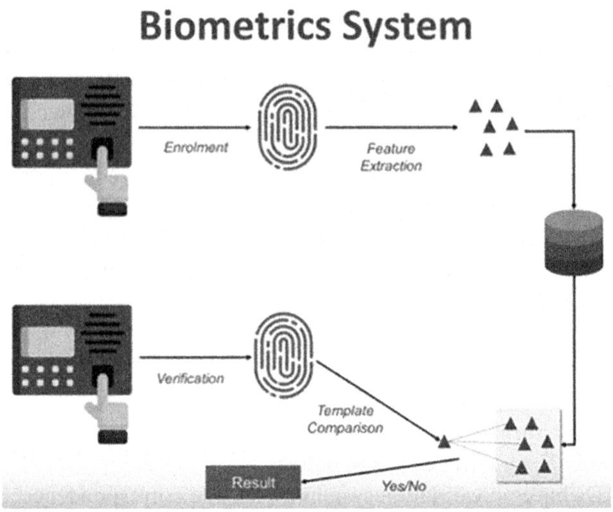

Biometrics System

FIGURE 9.5   Biometric system [6].

C. Individual User Level: IoT highlights in keen answers for homes, vehicles and other individual effects, and so forth.
D. Medical Services Organizations: Easy recovery and checking of the client information for better investigation of well-being measurements.

Figure 9.5 shows the biometric system for attendance using IoT.

9.7   CONCLUSION

In this chapter, we have discussed the architecture of IoT and given the significant safety issues of IoT safety assaults and their countermeasures. Because of the absence of security system in IoT gadgets, numerous IoT gadgets become vulnerable objectives and even this is not in the casualty's information on being contaminated. In this chapter, the security prerequisites are talked about, for example, privacy, honesty, and verification, and so forth. For security purposes in different applications, it is truly essential to introduce safety instrument in IoT and correspondence organizations. Additionally, to shield from any security attacks, it is likewise prescribed not to utilize default passwords for the gadgets and read the security prerequisites for the gadgets prior to utilizing it unexpectedly. Crippling the highlights that are not utilized may diminish the odds of security assaults. Additionally, it is imperative to examine diverse security conventions utilized in IoT gadgets and organizations.

REFERENCES

1. M. Abomhara and G. M. Koien, "Security and privacy in the Internet of Things: Current status and open issues," *Int'l Conference on Privacy and Security in Mobile Systems (PRISMS)*, pp. 1–8, 2014.

2. S. Chen, H. Xu, D. Liu, B. Hu, and H. Wang, "A vision of IoT: Applications, challenges, and opportunities with china perspective," *IEEE Internet of Things Journal*, vol. 1, no. 4, pp. 349–359, 2014.

3. L. Atzori, A. Iera, and G. Morabito, "The internet of things: A survey," *Computer Networks*, vol. 54, no. 15, pp. 2787–2805, 2010.

4. M. M. Hossain, M. Fotouhi, and R. Hasan, "Towards an analysis of security issues, challenges, and open problems in the internet of things," *Services (SERVICES), 2015 IEEE World Congress on. IEEE*, pp. 21–28, 2015.

5. K. Zhao and L. Ge, "A survey on the internet of things security," *Int'l Conf. on Computational Intelligence and Security (CIS)*, pp. 663–667, 2013.

6. L. Atzori, A. Iera, G. Morabito, and M. Nitti, "The social internet of things (SIoT) – when social networks meet the internet of things: Concept, architecture and network characterization," *Computer Networks*, vol. 56, pp. 3594–3608, 2012.

7. K. Kumar, E.S. Gupta and E.A.K. Rana, Wireless sensor networks: A review on "challenges and opportunities for the future world-LTE," *Amity Journal of Computational Sciences (AJCS)*, vol. 1, no. 2, pp. 30–34, 2018.

8. A.K. Rana, R. Krishna, S. Dhwan, S. Sharma and R. Gupta, "Review on artificial intelligence with internet of things-problems, challenges and opportunities," *2019 2nd International Conference on Power Energy, Environment and Intelligent Control (PEEIC)*, pp. 383–387. IEEE, 2019.

9. A.K. Rana and S. Sharma, "Contiki Cooja Security Solution (CCSS) with IPv6 routing protocol for low-power and lossy networks (RPL) in internet of things applications." in Nikhil Marriwala, C. C. Tripathi, Dinesh Kumar, and Shruti Jain (Eds.), *Mobile Radio Communications and 5G Networks,* pp. 251–259. Springer, Singapore, 2020.

10. M. Leo, F. Battisti, M. Carli, and A. Neri, "A federated architecture approach for Internet of Things security," *Euro Med Telco Conference (EMTC)*, pp. 1–5, 2014.

11. M. Farooq, M. Waseem, A. Khairi, and S. Mazhar, "A critical analysis on the security concerns of Internet of Things (IoT)," *Perception*, vol. 111, pp. 1–6, 2015.

12. R. Roman, P. Najera, and J. Lopez, "Securing the internet of things," *Computer*, vol. 44, pp. 51–58, 2011.

13. H. Ning, H. Liu, and L. T. Yang, "Cyberentity security in the internet of things," *Computer*, vol. 46, no. 4, pp. 46–53, 2013.

14. R. Roman, J. Zhou, and J. Lopez, "On the features and challenges of security and privacy in distributed internet of things," *Computer Networks*, vol. 57, pp. 2266–2279, 2013.

15. A. K. Rana, and S. Sharma, "Industry 4.0 manufacturing based on IoT, cloud computing, and big data: Manufacturing purpose scenario." In Gurdeep Singh Hura, Ashutosh Kumar Singh, Lau Siong Hoe (Eds.), *Advances in Communication and Computational Technology* (pp. 1109–1119). Springer, Singapore, 2019.

16. A. Kumar, A.O. Salau, S. Gupta and K. Paliwal, "Recent trends in IoT and its requisition with IoT built engineering: A review." In Banmali S. Rawat, Aditya Trivedi, Sanjeev Manhas, Vikram Karwal (Eds.), *Advances in Signal Processing and Communication* (pp. 15–25). Springer, Singapore, 2019.

17. A. K. Rana and S. Sharma, "Enhanced energy-efficient heterogeneous routing protocols in WSNs for IoT application," *IJEAT*, vol. 9, no. 1, pp. 4418–4425, 2019.

18. L. Da Xu, W. He, and S. Li, "Internet of things in industries: A survey," *IEEE Transactions on Industrial Informatics*, vol. 10, no. 4, pp. 2233–2243, 2014.

19. L. M. R. Tarouco, L. M. Bertholdo, L. Z. Granville, L. M. R. Arbiza, F. Carbone, M. Marotta, and J. J. C. de Santanna, "Internet of things in healthcare: Interoperability and security issues," *Communications (ICC), IEEE International Conference on. IEEE*, pp. 6121–6125, 2012.

20. A. Mohan, "Cyber security for personal medical devices internet of things," *Distributed Computing in Sensor Systems (DCOSS), 2014 IEEE International Conference on. IEEE*, pp. 372–374, 2014.
21. S. Yoon, H. Park, and H. S. Yoo, "Security issues on smarthome in IoT environment," in James J. (Jong Hyuk) Park, Ivan Stojmenovic, Hwa Young Jeong, and Gangman Yi (Eds.), *Computer Science and Its Applications* (pp. 691–696). Lecture Notes in Electrical Engineering, Springer, Guam, 2015.
22. S. Dhawan and R. Gupta, "Analysis of various data security techniques of steganography: A survey," *Information Security Journal: A Global Perspective*, vol. 30, no. 2, pp. 1–25, 2020.
23. S. Dhawan and R. Gupta, "Comparative analysis of domains of technical steganographic techniques," *2019 6th International Conference on Computing for Sustainable Global Development, 13–15* March, New Delhi, 2019.
24. Burak Kantarci, Melike Erol-Kantarci, and Stephanie Schuckers, "Towards secure cloud-centric internet of biometric things," *Cloud Networking (CloudNet), 2015 IEEE 4th International Conference on. IEEE*, pp. 81–83, 2015.
25. P. N. Mahalle, B. Anggorojati, N. R. Prasad, and R. Prasad, "Identity authentication and capability based access control (IACAC) for the internet of things," *Journal of Cyber Security and Mobility*, vol. 1, pp. 309–348, 2013.

10 Wireless Sensor Network for IoT-Based ECG Monitoring System Using NRF and LabVIEW

Ashish Gupta, Rajesh Kumar, and Devvrat Tyagi
North Eastern Regional Institute of Science and Technology

CONTENTS

10.1 INTRODUCTION

In recent years, the widely used analytical tool in cardiology is electrocardiography (ECG). Its contributions are considerable in the diagnostic and patient management having heart illness. Especially, it is useful in the diagnosis of the stern myocardial ischemic syndromes and cardiac arrhythmia. The signal processing is helpful in many heart diseases. For this reason it is important to acquire precise raw signal caused by heart muscle contraction and expansion, and further we can perform signal processing.

To measure the heart rate and regularity of heart ECG is very commonly used method. The count of heart beat and by seeing the waveform of ECG, doctors diagnose the irregularities of the heart. It is also helpful to diagnose the effect of valve condition, any effect of drug or device on heart ECG. The ECG wave consists of five systole and diastole named as P, Q, R, S and T [1]. The ECG wave is depicted in Figure 10.1. A portion of ECG waveform with peak R is known as QRS complex. The frequencies involved in the ECG are in the range of 0.05–100 Hz. The amplitude of signal is in the range of 1 to 10 mV. The ECG analyzing system performance is measured on the basis of reliability and accurate detection of the QRS complex in

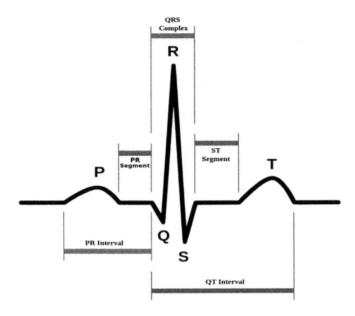

FIGURE 10.1   ECG waveform with PQRST marker [2].

the wave. The detection is possible using algorithms and the algorithms also help in the detection of T&P wave. The P-wave signifies the voltage generated when the atria, the upper compartments of the heart, gets activation from the nerve bundle node while QRS complex and T-wave in continuation represent the ventricular lower chamber excitation. The detection of QRS complex is important task in the analysis of ECG signal. After identification of QRS complex a detailed analysis is performed on the ECG data. The analysis includes calculating heart rate and ST segment [1,2]. The ECG is recorded with sensors lead placed at different locations on the chest as depicted in Figure 10.2, referred as lead positioning system.

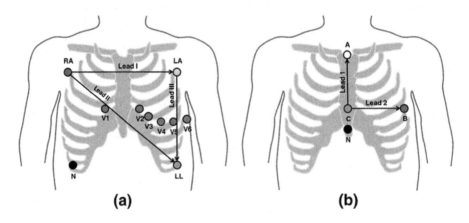

FIGURE 10.2   Lead positioning system for ECG recording [1].

Many physiological conditions of humans can be attained from the study of PQRST parameters recorded by an ECG recording instrument [3]. Application needs sure changes. For making timely changes virtual instrument support a lot. It offers user the flexibility of system design and powerful solution without including new machinery/ equipment or outdated instrument.

Real-time biomedical data can be recorded using smart devices. The system includes biomedical sensors, virtual instrument such as LabVIEW and national instrument hardware. Also the system reads the information from cloud storage such as Physio data bank by MIT-BIH for analysis. DAQ cards by national Instrument are also used to record data and connect sensor. The virtual instrument provides tool to validate and test biomedical instruments [4].

Some parameters are overlooked in many of the designed WSN nodes [6,7]. In Table 10.1, a comparative analysis of the proposed system with different existing is summarized. Some IoT systems have been designed for various application [9–12].

10.2   PROPOSED SYSTEM

In the proposed work we discussed a novel design of WSN node for low power consumption. The dimensions of the design are kept as small as possible. The node designed can act as sender or receiver, i.e., it is transceiver module. The sensor can be connected to the circuit easily and give output in analog or digital form over single pin. The block diagram of the system can be referred in Figure 10.3. The consumption

TABLE 10.1
Comparison of Proposed System with Existing System

Sl. No.	System	Features	Remarks
1	Patient health monitoring system using LabVIEW and wireless sensor network for IoT	Arduino board with Xbee and LabVIEW, sensor, GSM/GPRS (General Packet Radio Service), web publishing tool	Arduino board is used. No customization on PCB or circuit. Development boards and attachments used.
2	Health monitoring system based on IoT using Raspberry Pi and ECG Signal	Arduino Uno, Zigbee, Python language, SSL encryption	Raspberry pi is costly and needs Wi-Fi network for wireless operation. High power consumption.
3	Power optimization module in a network for WSN [8].	Arduino, XBee is used. Sleep mode implemented.	10% power consumption reduction is achieved. Range of transmission is less.

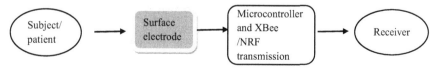

FIGURE 10.3   Block diagram of data acquisition system using LabVIEW.

varies based on different factor such as distance up to which data are to be transferred and the duration of data collection by the sensor. The maximum power consumption takes place during transmission. Power consumption is observed in two cases. Firstly, the sensor data are sampled at an interval of 1 min and then an interval of 15 s is taken for sampling. Comparing both the cases there is a drastic change in power consumption. Power consumption also gets to change during the communication protocol used in data transmission.

10.3   HARDWARE AND SOFTWARE DESCRIPTION

Sensing node has been made using Atmega-328 microcontroller IC from AVR (Alf and Vegard's Reduced Instruction Set Computer-RISC processor) family microcontroller. It has 28 pins in DIP package. The controller is an 8-bit microcontroller chip based on RISC architecture with picowatt power consumption. The controller has memory capabilities of 1024 B EEP Read Only Memory and 2 KB Static RAM. Twenty-three general-purpose Input/Output lines are helpful to interface input and output devices for small application such as data acquisition from human body, three timer/counters with compare modes, interrupts can be generated by programming internally and external events on pis are very useful in programming. The controller has a serial programming facility using USART, a 2-wire interface which is byte oriented and Serial Peripheral Interface port. For converting analog into digital, the Integrated Circuit has 6 channel A to D convertor of 10 bits. The operating voltage is 1.8 to 5.5 V.

NRF24101 is a trans-receiver on single chip which works on the frequency band of 2.4 to 2.4835 GHz ISM band. It requires 3.3 V and current of 15 mA. It interfaces with controller using SPI interface as shown in Figure 10.4. The data rate can be configured as 250 kbps, 1 Mbps or 2 Mbps. The frequency channel can be selected. NRF module can communicate with six other transceivers and up to five levels of tree topology network can be created. Each node is assigned with address in the octal format defined by 15-bit address.

LM35 integrated circuit temperature sensors are used in various applications from household body temperature measuring thermometers to automotive applications. They have advantages over traditional RTDs (Resistance Temperature Detector) or thermistors. They require no linearization or cold junction compensation. They provide better

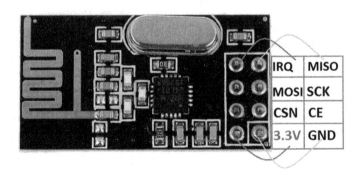

FIGURE 10.4   NRF24l01 module pin diagram.

noise immunity and the output can be obtained as logic signals that can be directly interface to digital systems. It works on the principle that the forward voltage of a silicon diode depends on its temperature and the relationship can be given by equation:

$$Vf = \frac{kT}{e} \ln \frac{I_f}{I_s} \text{ for } I_f \gg I_s$$

where T is the ambient temperature in degree kelvin, k is the Boltzmann's constant, e is the charge of electron, I_f is the forward current and I_s is the saturation current. I_s is a constant defined by the diode size. The voltage is either amplified for an analog output or fed to ADC to produce digital signal. The output of LM35 is linearly proportional to Celsius scale of temperature. The LM35 covers a temperature range of –55°C to 150°C. Its overall accuracy is ±0.75°C and near the room temperature is ±0.25°C. Internal offset produces 0 V at 0°C. This sensor has three pins to connect power, ground and the output as given in Figure 10.5. It draws 60 µA from the supply. The reason for choosing LM35 is summarized in Table 10.2.

ThingSpeak is a cloud platform service for IoT applications. It allows designer to store data on cloud and visualize the data. In commercial version it also provides analysis of live data stream in the cloud. It interacts with the IoT device using RESTful API and MQTT protocols.

LabVIEW software:

Virtual instrument has lot of advantages over traditional instruments. The software platform gives convenience to modify and add features as and when required. DAQ card facilitates the signal interfacing with computer and software display the signal on front panel. It also has LabVIEW VISA port and web publishing tools.

10.4 CIRCUIT DESCRIPTION

The system consists of two sections: transmission and receiver as shown in Figure 10.6. The sensor connected at transmitter section collects data from physical environment and converts the data in digital packets using microcontroller. The data packets are transferred to receiving section using NRF module.

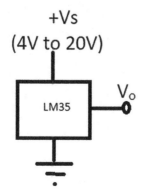

FIGURE 10.5 Temperature sensor.

TABLE 10.2
Advantages and Disadvantages of Widely Used Temperature Sensors

Device	Advantages	Disadvantages
RTD	Linear, high stability and wide range of operating temperature.	Slow response and low sensitivity. Requires three wire or four wire measurement. Sensitive to shock and vibration. Voltage source required.
Thermistor	High stability and fast response. High resistance eliminates the need of four-wire measurement. Small in size and interchangeable.	Non-linear and limited operating temperature. Voltage source required.
Thermocouple	Simple, wide range of operating temperature, no external power source required, rugged and inexpensive.	Non-linear, low stability, low sensitivity, low output voltage can be affected by Radio Interference and Electro Magnetic Interference. Cold junction compensation required.
IC-LM35	Linear, high sensitivity and inexpensive	Power supply required, subject to self-heating

a)   b)

FIGURE 10.6   Block diagram of the system. (a) Transmitter node. (b) Receiver node.

FIGURE 10.7   Image of sensing node with components mounted.

The sensing node circuit design and PCB schematics with components mounted are shown in Figure 10.7. The size of the PCB is approximately 5 cm in length and 4 cm in breadth. Height is around 3 cm.

10.5 WORKING OF PROPOSED SYSTEM

The central node, i.e., the receiver node, will process the data sent by sensing node. The program booted in the microcontroller is fetched from the memory whenever it is powered. The processing unit executes the command fetched sequentially as per the algorithm shown in Figure 10.7. After data acquisition using the sensor, the analog value converted into digital data is pushed onto the radio antenna of NRF24L01 module. For more than two nodes tree or star network topology is used. Each node is assigned an octal address like 00, 01, 92 and so on. The addressing is continued for the successive nodes. The logic for the programmed loaded in controller is shown in Figure 10.8 with the help of algorithm. The data is viewed in virtual instrumentation tool LabVIEW and the wiring diagram is shown in Figure 10.9 and the waveform developed is shown in Figure 10.10.

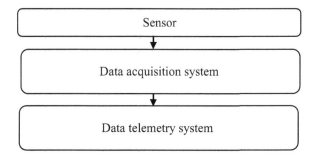

FIGURE 10.8 Algorithm for the sensing node.

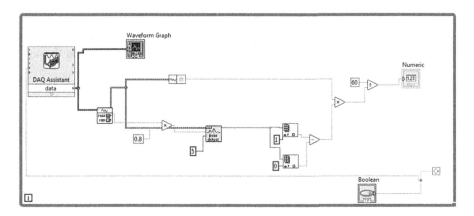

FIGURE 10.9 LabVIEW data assistant for collecting data from sensor.

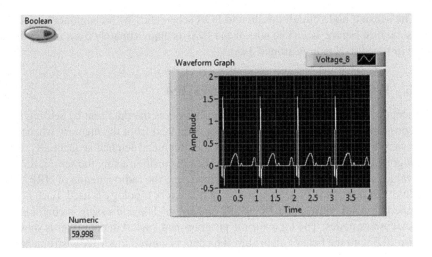

FIGURE 10.10 LabVIEW front panel to show ECG waveform after processing.

10.6 RESULTS

The experiment is designed with sensing node for transmission. One sample of data is transmitted using the node and the readings are taken for power consumption calculation. The amount of a charge stored in the battery is known as ampere-hour. A battery voltage decreases as the battery discharges; it does not give a perfect measure of how much energy is stored and we would need watt hour. To estimate the watt hour of the battery we take the product of average battery voltage with the battery volume in amp-hour. We will be using two battery cells for central node, one to power the NRF and microcontroller and the other to power the GSM (Global System for Mobile Communications).

$E = Cx V_{avg}$, where E is energy measured in watt-hours, C is for charge storage in amp-hours and V_{avg} is the average of voltage.

The comparison of different modules is summarized in Table 10.3. The temperature readings of LM35 are recorded as shown in Table 10.4 and the variation of temperature is depicted in graph as in Figure 10.11. The sensor was placed on the wrist of the subject. Room temperature was 25°C during the experiment setup.

TABLE 10.3

Comparison of Sensing Node Power Consumption Using Atmega328p on Customized Pcb, Arduino with NRF Module and NodeMCU with Wi-Fi

Parameter	Proposed Design	Arduino	Node-MCU
Voltage required for working	5 V	5 V	5 V
Current consumed	.03 A	.050 A	.080 A
Range	1,000 m	1,000 m	100 m

TABLE 10.4

Data Recorded from Temperature Sensor

Time	Temperature at Transmitter Side	Temperature at Receiver Side
0	33.75	33.75
1.2	33.94	33.94
2.4	34.06	34.06
3.5	34.13	34.13
4.8	34.19	34.19
5.5	34.25	34.25
5.8	34.31	34.31
6.5	34.88	34.88
7.2	35.0	35.0
8.1	35.56	35.56
9.1	36.1	36.1
10.3	36.4	36.4

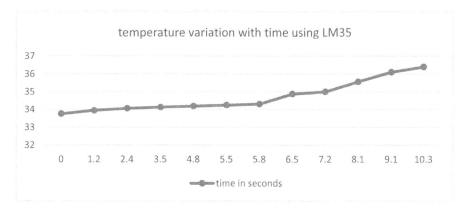

FIGURE 10.11   Temperature sensor data plot.

10.7 CONCLUSION

A WSN node has been designed for low power applications using microcontroller and NRF module. A network is established using tree or star topology for data monitoring. The deployment of nodes is for long duration and with renewable sources of energy it sustains for many days. The experimental results depict that the designed node consumes less power compared with previous design.

10.8 FUTURE SCOPE

To control the temperature we can implement PID control mechanism. PID control will automatically decrease the damping factor for system response and control process variables such as temperature and humidity [5,11]. In ICU the environmental parameters need to be precise for well-being of patients. The temperature in the environment is increased by a heater element. A motor driven fan near the heater draws in filtered fresh air. The health parameters data can be used for machine learning algorithm which might help to predict the condition of patient.

REFERENCES

1. Khanja P, Wattanasirichaigoon S, Natwichai J, Ramingwong L, Noimanee S, A WEB base system for ECG data transferred using ZIGBEE/IEEE technology, *Proceedings of the 3rd International Symposium on Biomedical Engineering (ISBME '08)*, Bangkok, Thailand, 2008.
2. Saritha C, Sukanya V, Narasimha Murthy Y, ECG signal analysis using wavelet transforms, *Bulgarian Journal of Physics*, 2008; 68–77.
3. zone.ni.com › Manuals › LabVIEW 2013 Biomedical Toolkit Help.
4. Biomedical User Group Forum; http://decible.ni.com/content/groups/biomedicalusergroup.
5. Bansal H, Dr. Mathew L. Ashish Gupta controlling of temperature and humidity for an infant incubator using microcontroller, *International Journal of Advanced Research in Electrical, Electronics and Instrumentation Engineering*, 2016; 4(6): 11–15.
6. Gungor VC, Lu B, Hancke GP. Opportunities and challenges of wireless sensor networks in smart grid. *IEEE Transactions on Industrial Electronics*, 2010; 57(10): 3557–3564.
7. Meddour F, Dibi Z. An efficient small size electromagnetic energy harvesting sensor for low-DC-power applications. *IET Microwaves, Antennas & Propagation*, 2016; 11(4): 483–489.
8. Suryadevara NK, Mukhopadhyay SC, Kelly SD, Gill SP. WSN-based smart sensors and actuator for power management in intelligent buildings. *IEEE/ASME Transactions on Mechatronics*, 2014; 20(2): 564–571.
9. Gupta A, Kumar R. An IOT enabled air quality measurement. *Indian Journal of Science and Technology*, 2018; 11(46). DOI: 10.17485/ijst/2018/v11i46/139720.
10. Kumar R, Jain P, Ashishgupta. Internet of things enable home appliances control and monitoring system. *Smart Home Control and Monitoring System based on Internet of Things with AI*, ICAITCT, 2016; 10(6): 24–30.
11. Mittal H, Mathew L, and Gupta A. Design and development of an infant incubator for controlling multiple parameters. *International Journal of Emerging Trends in Electrical and Electronics*, 2015; 11(5): 32–39.
12. Gautam A, Verma G, Qamar S, Shekhar S. Vehicle pollution monitoring, control and challan system using MQ2 sensor based on internet of things. *Wireless Personal Communications*, 2073–2075, https://doi.org/10.1007/s11277-019-06936-4, 2019.
13. https://electronicshobbyists.com/nrf24l01-interfacing-with-arduino-wireless-communication

11 Towards Secure Deployment on the Internet of Robotic Things: Architecture, Applications, and Challenges

Arun Kumar Rana and Sharad Sharma
Maharishi Markandeshwar (Deemed to be University)

Sachin Dhawan
Panipat Institute of Engineering and Technology

Shubham Tayal
SR University

CONTENTS

DOI: 10.1201/9781003138068-11

11.1 INTRODUCTION

The Internet of Robotic Things (IoRT), the confluence of the Internet of Things (IoT) and robotics, is a notion in which autonomous machines collect and interact with information from various sensors embedded and sourced to perform tasks requiring critical thinking. IoT is a vibrant and active area of research [1], and at the same time, robotics is a strong and well-established field with many applications [2–4]. British innovator Kevin Ashton, who first discovered IoT in 1999, and Czech author Karel Capek first used the term robot from the Slavic word 'robota' meaning 'forced labor' in his play RUR, combining the two terms that our new kid gets in the industry is IoT as seen in Figure 11.1. As it happens in the concept of IoT, the everyday devices or "things" that usually do not have internet is powered by the internet and related technologies such as cloud computing. Similarly in IoRT, the 'things' are Robots that are used for industrial applications and are now connected to various networks such as the internet and can move their data to sophisticated cloud

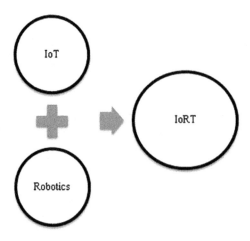

FIGURE 11.1 Internet of robotic things.

platforms. IoT, the advances, structures, and administrations that permit enormous quantities of sensor-empowered, remarkably addressable "things" to speak with one another and move information over unavoidable systems utilizing Internet conventions, is required to be the following incredible mechanical development and business opportunity [19–22]. It will surpass in size and significance both the PC and portable interchanges showcase, and even the advancement of the Internet itself. As of now, most IoT operations are cantered by the use of related gadgets with easy, locally accessible, inactive sensors to track, monitor, and advance frameworks and their procedures. This will be massively productive on its own in any case; it is not too early to explore the further evolved and revolutionary components of universal accessibility to and communication between, keen gadgets for ground-breaking organizations. This exploration study presents the idea of the IoRT, where shrewd gadgets can screen occasions, intertwine sensor information from an assortment of sources, utilize nearby and disseminated "insight" to decide the best strategy, and afterward act to control or control protests in the physical world, and at times while truly traveling through that world. It will likewise look at the numerous ways IoT innovations and automated "gadgets" cross to give progressed mechanical capacities, alongside novel applications, and by expansion, new business, and venture openings. IoRT unites self-ruling automated frameworks with the IoT vision of sensors and savvy questions inescapably installed in regular situations [5–8]. This consolidation can empower novel applications in pretty much every division where collaboration among robots and IoT innovation can be envisioned.

11.1.1 MOTIVATION AND CONTRIBUTION

The remarkable recent developments in IoT with robotic technology have provided the opportunity to broadly deploy tiny sensors for wireless communication for various applications like smart homes, small cities, smart healthcare and smart industry [23,24]. The objective of the paper provides an overview of the integration of Robotics with the IoT. This paper's contribution can be summarized as follows:

- Discussing recent articles that investigate robotic integration with different IoT applications.
- Investigating IoRT challenges and how the integration of IoT with robotic will solve them.
- Providing different IoRT applications that benefit from fog integration with the IoT.
- Discussing the challenges, the integration of IoT with robotic poses.

The rest of the article is organized as follows: In Section 11.2 literature survey is addressed based on IoRT. Section 11.3 discussed IoRT and discusses emerging trends of IoRT. Section 11.5 addressed system architecture based on the different layers. Section 11.6 addressed the applications of IoRT. Section 11.7 discussed the research challenges of IoRT. The article is eventually summarized in Section 11.8. Moreover, the outline of the paper has been illustrated in Figure 11.2.

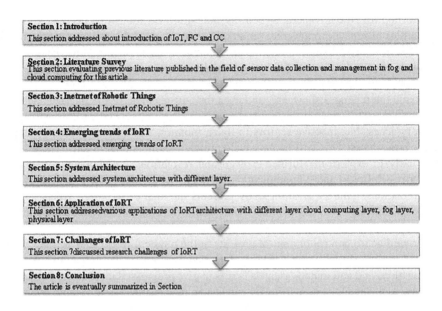

Section 1: Introduction
This section addressed about introduction of IoT, FC and CC

Section 2: Literature Survey
This section evaluating previous literature published in the field of sensor data collection and management in fog and cloud computing for this article

Section 3: Inetrnet of Robotic Things
This section addressed Inetrnet of Robotic Things

Section 4: Emerging trends of IoRT
This section addressed emerging trends of IoRT

Section 5: System Architecture
This section addressed system architecture with different layer.

Section 6: Application of IoRT
This section addressed various applications of IoRT architecture with different layer cloud computing layer, fog layer, physical layer

Section 7: Challanges of IoRT
This section ?discussed research challenges of IoRT

Section 8: Conclusion
The article is eventually summarized in Section

FIGURE 11.2   Outline of the paper.

11.2   LITERATURE SURVEY

The IoT brings a digital heartbeat to physical objects. And robotics is a branch of computer science and engineering that deals with machines that can work autonomously. And what happens when these two technologies unite? IoRT is a concept where IoT data help machines interact with each other and take required actions. In simpler words, it refers to robots that communicate with other robots and take appropriate decisions on their own. Pervasive sensors, cameras and actuators embedded in the surroundings and also self-help robots collect information in real-time. In the article "Flexible IoT Middleware for Integration of Things and Applications" [12], Bowman believes that while the IoT is far from all-embracing, every day is closer. In his paper "Context-Aware Computing The IoT: A Survey" in 2014 Charith Perera emphasizes that the number of sensors operating in the world is rising at an increasing rate as we move toward the IoT. Market reports have demonstrated substantial growth of new sensors over the past decade and predicting an increasing growth rate in the future. These sensors produce a large volume of data continuously. Zhen Peng—2012—claims that more and more network-based applications are being related to the Internet in his article entitled "Message-oriented Middleware data processing model in the Internet of Things" [13]. Devices have created several diverse types of applications; however, the maintenance and operation of the network still have new problems. In his article "Proposal for a Stable, Deployable and Transparent Middleware on the Internet of Things," [14] Hiro Gabriel Cerqueira Ferreira—2014—proposes a security framework for IoT middleware that is built such that real-life artifacts are converted into the virtuoso universe. In an earlier [15] paper Hiro suggested an IoT middleware and extensive usage scenarios, which were based on existing technologies in the article above when describing a security architecture. Javardhana Gubbi (2013) discussing the role of all-embracing sensing enabled by wireless sensor network (WSN) technology in many areas of modern life in his article "Internet of Things (IoT): a vision, design features, and potential directions." Chengjia Huo offers a middleware program capable of providing more reliable security by connecting the IoT to a server with the 'IoT Cloud Conversion across Application Domains' article [17].

11.3   INTERNET OF ROBOTIC THINGS

The integration of the Internet and RFID, sensors and intelligent objects is IoT. IoT enables individuals and objects to be connected with something or anything at any time, everywhere, ideally using any route and any service. This relation between the 'things' is shown in Figure 11.3. In three paradigms, IoT can be achieved: middleware, sensors, and the Knowledge Base, which communicate with each other and satisfy the Internet for Things' driven visions. The entire communication vision cannot be fulfilled by one paradigm [25]. The primary goal for creating and linking objects on the network would be the intersection of these visions.

The stuff that will be active participants are business, knowledge, and social processes for every network. The wireless sensor has introduced technical advancements for circuits and communications in the hardware realm, offering powerful and durable devices for sensing. This has contributed to the use of wireless networking systems in different settings. Sensor data are processed and sent to a centralized, distributed or hybrid processing module of the processing data. Therefore, a WSN needs

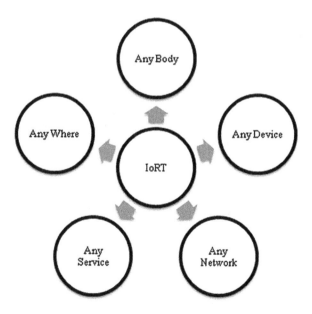

FIGURE 11.3   Internet of robotic things.

to meet many obstacles to building a viable IoT communication network. The IoT, the interconnection and communication between ordinary devices make several applications in many fields. Many of the characteristics that can and should be considered when designing an application apply to the availability of the network, latency, area of coverage, redundancy, user engagement and analysis of effect [26].

IoRT can be the perfect choice for industries that deal with heavy-duty work or repetitive manual jobs. Let us check out a few potential use cases through which industries can benefit from this newly emerged concept.

- Robots at warehouses can inspect product quality, check for product damages, and also help with put away. Without humans playing any role, robots can analyze the surroundings with the IoT data and respond to situations as needed.
- A robot can effectively play the role of a guidance officer and help customers with parking space availability. By checking the parking lots, robots can assist customers with the right place to park their vehicles.
- Robots can automate the labor-intensive and life-threatening jobs at a construction site. Right from scaffolding to loading and unloading heavy construction equipment, robots can take care of every on-site task responsibly. With the help of intelligent robots, construction engineers and managers can ensure enhanced worker health and safety.

However, according to the IoRT principle, all robots whose tasks are improved by the capabilities of the smart world and the smart environment benefit from shared integration, in which agents (robots) appear in addition to controlling functions and executing simple acts with simple mechanisms inside the Smart World, to execute complex operations. Robots will obtain tasks from the Smart Environment as a part of this integration, and also tracks the progress of mission execution and provides feedback to robots from the Smart Environment sensor network, for example, for optimum navigation, avoidance of obstacles/collisions, or successful contact between humans and robots. The IoRT is also a more sophisticated level of the IoT, which enables new technology such as cloud storage, wireless sensing and actuating, data processing, information support, guidance, surveillance, and security control, distributed tracking and networking to be implemented.

11.4 EMERGING IORT TECHNOLOGIES

The "things" are heterogeneous and are incorporated into various platforms, with different levels of sophistication and sensory/actuating, communication, processing, intelligence, and mobility [27].

11.4.1 SENSORS AND ACTUATORS

The two IoT standard inventions and their autonomy, characterized and recognized in all areas, are sensor gadgets and actuators; these are both permanently main segments for up-to-date IoRT frames for all types of interfaces and for the provision of these features to the IoRT stage using cooperative parts. Not as valuable utility is achieved in IoRT squares as the IoT sensors and actuators.

11.4.2 COMMUNICATION TECHNOLOGIES

The correspondence engineering of IoRT needs new methodologies empowering shared continuous calculation and the trading of information streams (vital for 3D-mindfulness and vision system) joined with inside correspondence, and edge figuring to empower the virtualization of capacities on the current processing motors while empowering the usability of such foundations in numerous areas. The correspondence framework and the IoRT outer correspondence should have the option to perform time-basic correspondence to guarantee impact counteraction gets conceivable, along these lines vigorously decreasing mishaps and crashes.

11.4.3 CONNECTED ROBOTIC THINGS

Fusion Information Robot things connected to them can share, intertwine and describe their sensor information. The versatility and independence abilities of automated get the issue of sensor combination IoT stages to a completely new degree of unpredictability and include altogether additional opportunities. Intricacy is expanded due to the extraordinary sum and assortment of sensor information that mechanical things can give and because the area of the detecting gadgets is not fixed and regularly is not known with conviction. Additional opportunities are empowered as a result of the capacity of mechanical things to self-sufficiently move to explicit areas to gather explicit tangible info, in light of the investigation of the as of now accessible information and of the displaying and thinking objectives [28].

11.4.4 VIRTUAL AND AUGMENTED REALITY

Robot-assisted frameworks in surgical treatment are technologies that include creativity in the workplace in virtual reality (VR) and in augmented reality (AR). Live and virtual imagery on robot-helped interfacing allows the mechanical instrument expert to track and speaks to an open stage for VR and AR capacity expansion. Live careful imaging is used to develop medical frameworks through image infusion or overlap of explicit objects with the robot-helped procedure. In a robot-assisted medical practice, the VR/AR invention is used to trace the movement of automated instruments in a smart patient life model structure seen on a comfort monitor.

11.4.5 VOICE RECOGNITION AND VOICE CONTROL

Today, chatbots and driven mouthpiece gadgets canter conversational interfaces. The promotion of IoT technologies and computerized work requires an extended variety of endpoints to connect people and mechanical things. As the IoRT work develops, the agreeable connection between mechanical things rises, making the structure for another nonstop and encompassing computerized experience where automated things and people are teaming up.

11.5 ARCHITECTURE OF IORT

The architecture of the IoRT can be divided into five layers, namely, (i) the hardware/ robotic things layer: comprising physical components such as the robots and its sensors; (ii) the network layer: consisting of connectivity options such as WiFi, ZigBee, etc.; (iii) the internet layer: this being the most crucial layer of the architecture here various energy-efficient and robust communication protocols such as MQTT (Message Queuing Telemetry Transport), IP (Internet Protocol); (iv) infrastructure layer: in this layer, the cloud computing platform and various machine learning tools come into the picture. These can be integrated with various robotic tools such as ROS (Robotic Operating System) here; (v) the application layer: this is the topmost layer of the architecture which helps in widening the user experience and here the end-user of the robot gets to see the desired outcome of the IoRT implementation on the robot, as shown in Figure 11.4.

11.5.1 HARDWARE LAYER

This is the lower layer of many robotic products, including vehicles, cameras, laptops, safeguards, aquatic robots, temperature sensors, personal appliances, home appliances and industrial sensors. Theoretically, concrete (real-life equivalent) covers that layer of abstraction to take advantage of the above-mentioned layer, i.e., the network layer.

11.5.2 NETWORK LAYER

Many network access solutions are provided in this second lower layer. Wireless networking is also enabled, such as 3G and LTE/4G. Few short-range connectivity technologies, such as WiFi, Bluetooth Low Energy and 6LoWPAN (BLE), the BGAN (Broad-Band Global Area Network), and Near Field Communication (NFC), have been included to facilitate less smooth networking among nearby robotic networks.

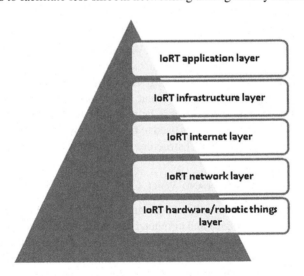

FIGURE 11.4 Architecture of internet of robotic things.

11.5.3 INTERNET LAYER

The central element of all communication is in IoRT architecture. For energy-efficient resource constraint and lightweight information retrieval on robotic systems due to their virtue, IoT special communication protocols were implemented on this layer selectively. MQTT, CoAP, XMPP, IPv6, UDP and DDS protocols support these functions: publishing/subscription communications, multi-CAST support in real-time, packet-switched networking, TCP alternative, networked embed system sharing, data map protection, middleware environment message queuing, fast local automation and direct addressing.

11.5.4 INFRASTRUCTURE LAYER

This part of the infrastructure is redesigned by the IoT-based robotic cloud stack to be the most useful system of all (cloud, middleware, business method and big data service-centered approaches). Data out of many collections and studies, and system control, synchronization of map cum weather data, and accumulation of sensor data are of most significance. Some scenarios of IoRT apply to a connected factory wherein the Robot data are monitored in real-time in the cloud as well as the maintenance of the robot is and performance optimization takes place well in advance. Another scenario which has been implemented by some companies is the concept where the Robot data from the factory are synced in real-time with the cloud and the actuation happens accordingly so that the productivity is high and no robot remains idle in the production line; this way it is ensured that the process of manufacturing never comes to a halt even because of the slightest of the error or faults in the robot. One of the most challenging scenarios in the industry is to develop a software platform, which is independent of the Robot being connected to it, i.e., different robots but all connected to one software platform on the cloud; this is the user experience and remote monitoring solutions become simplified to leaps and bounds in this [29].

11.5.5 APPLICATION LAYER

This is the highest layer of the IoRT architecture, which can disseminate user experience by analyzing the sample of available software using robotics. IoT-bound robots are dedicated to coping with different problems, for example, health systems, facilities management, convenience stores, critical living environments, data centers, exhibits on the markets, and more. Infinite and ever-growing prospects remain, hence their value and life [30].

11.6 APPLICATIONS OF IORT

In different fields of our lives, it is possible to incorporate robots into smart environments: home automation, wellness, transportation and logistics (see Figure 11.5). We will incorporate Smart- and Software-Based Buildings as a concept for smart environments. These are programmable buildings where sensing is implemented to execute different tasks, such as presence tracking, behavior and identity identification,

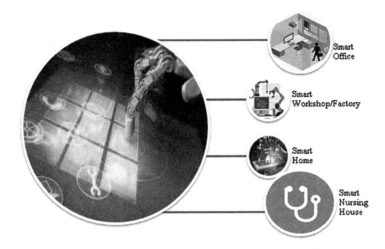

FIGURE 11.5   Applications of IoRT.

and user emotional state detection based on hardware and software. With different features, the sensing IoT functions can be introduced with different hardware modules, such as detectors for human occupancy, sensors for orientation and monitoring, etc. Let us address IoRT's program instances.

11.6.1   SMART HOME

To help sustain humans, the Smart House should be provided with enough sensors and apps to communicate with people and robots. Smart domestic programs can include food surveillance with food tips, children's homework assistance, special disability therapy surveillance, etc. The robots are designed to assist in household activities [31].

11.6.2   SMART OFFICE

The Smart Office can be supplied with sensors and software for the guest reception, meeting service, and various workflow facility, and monitor an optimized micro place. Simultaneously robots were designed to communicate easily with both guests and employees [32].

11.6.3   SMART WORKSHOP/FACTORY

To assist the production process, Smart Workshop/Factory can track and manage manufacturing/processing and accounting, manufacturing and quality control items, equipment construction, emergency monitoring and environmental adjustment needs. Simultaneous loading/unloading, manufacture (soldering, assembly, etc.) of several robots can be used, as well as quality control in the industrial area, while even mobile robots can control machines and machinery that are unique to humans [33].

11.6.4 SMART NURSING HOUSE

For patients with different diagnoses, Smart Nursing House offers unique person-oriented care. IoRT programs should be combined to reduce the expense of repetitive tasks, as some patients need around-the-clock treatment. The Activity Announcing Service will announce planned activities every morning. Events and news and the Support Robot may be tailored to a patient's particular work [34]. The maintenance workers may be partly replaced by IoRT, and call for assistance if necessary. A wandering elderly individual can be identified by the Smart environment, identify behavioral abnormalities, and those entities should be able to support a service robot. In the patient's room, numerous sensors and instruments must be mounted for tracking, taking drugs, and controlling environmental conditions [35–37].

11.7 RESEARCH CHALLENGES

Research on the IoRT is in its preliminary stage, being a novel field. Many problems must be discussed in detail by the scientific community (see Figure 11.6). This chapter outlines some of the daunting IoRT research problems that will be followed by a few potential applications.

11.7.1 SECURITY

In this area, security is a complex and challenging problem, related both to the security of the IoT and to the security of connecting robots. Any of the following factors listed below are responsible for major cybersecurity issues in robotics: Insecure communication between users and robots leads to cyber-attacks. In no time, hackers can easily hack into vulnerable ties of communication. In no time, hackers can easily hack into vulnerable communication links [9].

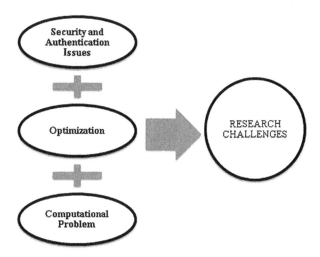

FIGURE 11.6 Research challenges in IoRT.

11.7.2 Authentication Issues

Failure to protect against unauthorized access will easily allow hackers, without using any legitimate username and password, to penetrate the robot systems and use their functions from remote locations. The lack of proper encryption on the part of vendors will expose potential hackers to sensitive data. Most features of the robot are programmable and affordable. If the default configuration of the robot is poor for hacking, it is easy for intruders to reach and modify the programmable features [10].

11.7.3 Computational Problem

One of the key advantages of IoRT is the capacity of the mutual stacking of computationally concentrated undertakings to the IoT cloud for execution. In any case, the choice to shared-of the stacking of a particular errand requires an increasingly stringent and joined engineering system that can deal with an assortment of complex issues [11].

11.7.4 Optimization

The computational test would deteriorate on the off chance that we do not think about advancement. Typically, the handling of assignment through-loading is settled on three execution techniques, including independent calculation by the individual mechanical framework, synergistic calculation by a gathering of the automated framework associated through a system and cloud calculation [18]. Here and there, a crossover cloud model incorporates fractional calculation taking every one of these systems together.

11.8 CONCLUSION

As the name implies, the IoRT is the amalgamation of two cutting-edge technologies, the IoT and Robotics. The vision behind this concept is to empower a robot with intelligence to execute critical tasks by itself. To comprehend this technology better, let us first break it down into its components. Internet-enabled robots represent the next frontier of IoT technologies and will become very popular shortly. Thus, the IoRT is becoming an important discipline in the natural course of IoT development. Additionally, with this new term, it is worth noting that many familiar concepts are having their own IoT spin. For example, there is an Internet of Battlefield Things, and, even an Internet of Prisons. This goes to show how IoT will be a defining trend in the years to come and sweep everything in its path. In this paper research challenges and emerging trends are presented in a precise manner so that enthusiasts can get involved in this novel concept in recent future.

REFERENCES

1. N. Alam, P. Vats, and N. Kashyap. (2017, October). Internet of Things: A Literature Review. In *2017 Recent Developments in Control, Automation Power Engineering (RDCAPE)*, pp. 192–197.
2. L. Royakkers and R. van Est. (2015). A Literature Review on New Robotics: Automation from Love to War. *International Journal of Social Robotics*, 7, 549–570.

3. B. Preising, T. C. Hsia, and B. Mittelstadt. (1991). A Literature Review: Robots in Medicine. *IEEE Engineering in Medicine and Biology Magazine*, *10*, 13–22.
4. N. Mohamed, J. Al-Jaroodi, and I. Jawhar. (2019). Fog-enabled Multi-robot Systems. In *2018 IEEE 2nd International Conference on Fog and Edge Computing (ICFEC)*, pp. 1–10. IEEE. https://www.overleaf.com/6319157491pnzdmdkhxtbv
5. A. K. Rana, and S. Sharma. (2021). Industry 4.0 Manufacturing Based on IoT, Cloud Computing, and Big Data: Manufacturing Purpose Scenario. In *Advances in Communication and Computational Technology*, pp. 1109–1119. Springer, Singapore.
6. A. K. Rana, and S. Sharma. (2021). Contiki Cooja Security Solution (CCSS) with IPv6 Routing Protocol for Low-Power and Lossy Networks (RPL) in Internet of Things Applications. In *Mobile Radio Communications and 5G Networks,* pp. 251–259. Springer, Singapore.
7. M. Dragone. (2017). Combining IoT, Robotics and AI: Where is the Added Value, Where are the Challenges. In *Workshop at the European Robotics Forum 2017*, Edinburgh, Scotland, UK, March 24.
8. M. Dragone. (2017). Combining IoT and Intelligent Robotics, Challenges and Opportunities. In *Workshop at IoT Week 2017*, Geneva, June 7.
9. B. Kehoe, S. Patil, P. Abbeel, and K. Goldberg. (2015). A Survey of Research on Cloud Robotics and Automation. *IEEE Transactions on Automation Science and Engineering*, *12*(2), 398–409.
10. N. Dragoni, A. Giaretta, and M. Mazzara. (2016). The Internet of Hackable Things. In *International Conference in Software Engineering for Defence Applications,* pp. 129–140. Springer, Cham.
11. L. A. Grieco, A. Rizzo, S. Colucci, S. Sicari, G. Piro et al. (2014). IoT-aided Robotics Applications: Technological Implications, Target Domains and Open Issues. *Computer Communications*, *54*, 32–47.
12. M. A. Chaqfeh, and N. Mohamed. (2012). Challenges in middleware solutions for the internet of things. In *2012 international conference on collaboration technologies and systems (CTS)*, pp. 21–26. IEEE.
13. J. Boman, J. Taylor, and A. Ngu. (2014). Flexible IoT Middleware for Integration of Things and Applications. In *Proc. 10th IEEE Int. Conf. Collab. Comput. Networking, Appl. Work.*, no. CollaborateCom pp. 481–488.
14. Z. Peng, Z. Jingling, and L. Qing. (2012). Message-oriented Middleware data processing model in internet of things. In *Proc. 2012 2nd Int. Conf. Comput. Sci. Netw. Technol.*, pp. 94–97.
15. H. Gabriel, C. Ferreira, R. Timóteo, D. S. Júnior, F. Elias, and G. De Deus. (2014). Proposal of a Secure, Deployable and Transparent Middleware for Internet of Things. Information Systems and Technologies (CISTI).
16. H. G. C. Ferreira, E. D. Canedo, and R. T. Sousa Júnior. (2013). IoT Architecture to Enable Intercommunication Through REST API and UPnP Using IP, ZigBee and Arduino. In *2013 IEEE 9th International Conference on Wireless and Mobile Computing, Networking and Communications (WiMob)*, pp. 53, 60.
17. D. Miorandi, S. Sicari, F. De Pellegrini, and I. Chlamtac. (2012). Internet of Things: Vision, Applications and Research Challenges. *Ad Hoc Networks*, *10*(7), 1497–1516.
18. G. Hu, W. P. Tay, and Y. Wen. (2012). Cloud Robotics: Architecture, Challenges and Applications. *IEEE Networks*, *26*(3), 21–28.
19. A. Kumar, A. O. Salau, S. Gupta, and K. Paliwal. (2019). Recent Trends in IoT and Its Requisition with IoT Built Engineering: A Review. In *Advances in Signal Processing and Communication* (pp. 15–25). Springer, Singapore.
20. Rana, A.K. and Sharma, S. (2019). Enhanced energy-efficient heterogeneous routing protocols in WSNs for IoT application. *International Journal of Engineering and Advanced Technology* (IJEAT) , *9*(1).

21. K. Kumar, E. S. Gupta, and E. A. K. Rana. (2018). Wireless Sensor Networks: A Review on "Challenges and Opportunities for the Future World-LTE". *Amity Journal of Computational Sciences (AJCS), 1*(2).

22. A. K. Rana, R. Krishna, S. Dhwan, S. Sharma, and R. Gupta. (2019, October). Review on Artificial Intelligence with the Internet of Things-Problems, Challenges, and Opportunities. In *2019 2nd International Conference on Power Energy, Environment and Intelligent Control (PEEIC)*, pp. 383–387. IEEE.

23. D. Zhu, X. Feng, X. Xu, Z. Yang, W. Li, S. Yan, and H. Ding. (2020). Robotic Grinding of Complex Components: A Step Towards Efficient and Intelligent Machining–Challenges, Solutions, and Applications. *Robotics and Computer-Integrated Manufacturing, 65*, 101908.

24. A. Soni and A. Rana. (2017). Analyze Portrayal of Stable Election Protocol for Wireless Sensor Network using Matlab [C]. *International Journal of Computer Applications, 174*(8), 18–22.

25. L. Romeo, A. Petitti, R. Marani, and A. Milella. (2020). Internet of Robotic Things in Smart Domains: Applications and Challenges. *Sensors, 20*(12), 3355.

26. A. K. Rana, A. Salau, S. Gupta and S. Arora. (2018). A Survey of Machine Learning Methods for IoT and Their Future Applications. *Amity Journal of Computational Sciences, 2*(2), 1–5.

27. A. Banerjee, C. Chakraborty, A. Kumar, and D. Biswas. (2020). Emerging Trends in IoT and Big Data Analytics for Biomedical and Health Care Technologies. In *Handbook of Data Science Approaches for Biomedical Engineering*, pp. 121–152. Academic Press.

28. A. Banerjee, C. Chakraborty, and Sr, M. Rathi. (2020). Medical Imaging, Artificial Intelligence, Internet of Things, Wearable Devices in Terahertz Healthcare Technologies. In *Terahertz Biomedical and Healthcare Technologies*, pp. 145–165. Elsevier.

29. L. Romeo, A. Petitti, R. Marani, and A. Milella. (2020). Internet of Robotic Things in Smart Domains: Applications and Challenges. *Sensors, 20*(12), 3355.

30. Y. Masuda, A. Zimmermann, S. Shirasaka, and O. Nakamura. Internet of Robotic Things with Digital Platforms: Digitization of Robotics Enterprise. In *Human Centred Intelligent Systems*, pp. 381–391. Springer, Singapore.

31. L. Romeo, A. Petitti, R. Marani, and A. Milella. (2020). Internet of Robotic Things in Smart Domains: Applications and Challenges. *Sensors, 20*(12), 3355.

32. Y. Liu, W. Zhang, S. Pan, Y. Li, and Y. Chen. (2020). Analyzing the Robotic Behavior in a Smart City with Deep Enforcement and Imitation Learning Using IoRT. *Computer Communications, 150*, 346–356.

33. G. C. Deac, C. N. Georgescu, C. L. Popa, and C. E. Cotet. (2020). Virtual Reality Digital Twin for a Smart Factory. *International Journal of Modeling and Optimization, 10*(6) 220–228.

34. W. I. Liu. (2020). New Opportunities for Healthcare Driven by Smart Technology. *Hu Li Za Zhi, 67*(5), 4–5.

35. A. K. Rana, and S. Sharma. (2021). The Fusion of Blockchain and IoT Technologies with Industry 4.0. In *Intelligent Communication and Automation Systems*, pp. 275–290. CRC Press.

36. S. Dhawan, R. Gupta, A. Rana, and S. Sharma. (2021). Various Swarm Optimization Algorithms: Review, Challenges, and Opportunities. In *Soft Computing for Intelligent Systems*, pp. 291–301. Springer.

37. A. Kumar and S. Sharma. (2021). IFTTT Rely Based a Semantic Web Approach to Simplifying Trigger-Action Programming for End-User Application with IoT Applications. In *Semantic IoT: Theory and Applications: Interoperability, Provenance and Beyond*, pp. 385–397.

Index

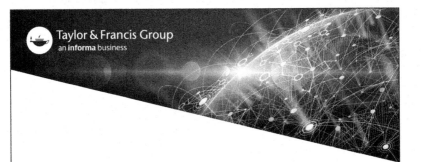

Printed in the United States
by Baker & Taylor Publisher Services